Mathematical and Computational Techniques for Multilevel Adaptive Methods

Frontiers in Applied Mathematics

Frontiers in Applied Mathematics is a series that presents new mathematical or computational approaches to significant scientific problems. Beginning with Volume 4, the series reflects a change in both philosophy and format. Each volume focuses on a broad application of general interest to applied mathematicians as well as engineers and other scientists.

This unique series will advance the development of applied mathematics through the rapid publication of short, inexpensive books that lie on the cutting edge of research.

Frontiers in Applied Mathematics

Vol. 1	Ewing, Richard E., The Mathematics of Reservoir Simulation
Vol. 2	Buckmaster, John D., The Mathematics of Combustion
Vol. 3	McCormick, Stephen F., Multigrid Methods
Vol. 4	Coleman, Thomas F. and Van Loan, Charles, Handbook for Matrix Computations
Vol. 5	Grossman, Robert, Symbolic Computation: *Applications to Scientific Computing*
Vol. 6	McCormick, Stephen F., Multilevel Adaptive Methods for Partial Differential Equations
Vol. 7	Bank, R. E., PLTMG: A Software Package for Solving Elliptic Partial Differential Equations. *Users' Guide 6.0*
Vol. 8	Castillo, José E., Mathematical Aspects of Numerical Grid Generation
Vol. 9	Van Huffel, Sabine and Vandewalle, Joos, The Total Least Squares Problem: *Computational Aspects and Analysis*
Vol. 10	Van Loan, Charles, Computational Frameworks for the Fast Fourier Transform
Vol. 11	Banks, H.T. , Control and Estimation in Distributed Parameter Systems
Vol. 12	Cook, L. Pamela, Transonic Aerodynamics: *Problems in Asymptotic Theory*
Vol. 13	Rüde, Ulrich, Mathematical and Computational Techniques for Multilevel Adaptive Methods

Mathematical and Computational Techniques for Multilevel Adaptive Methods

Ulrich Rüde
Technische Universität München

Society for Industrial and Applied Mathematics
siam
Philadelphia 1993

Library of Congress Cataloging-in-Publication Data

Rüde, Ulrich
Mathematical and computational techniques for multilevel adaptive methods \ Ulrich Rüde.
p. cm. — (Frontiers in applied mathematics ; vol. 13)
Includes bibliographical references and index.
ISBN 0-89871-320-X
1. Differential equations, Partial—Numerical solutions.
2. Multigrid methods (Numerical analysis) I. Title. II. Series: Frontiers in applied mathematics ; 13.
QA377.R87 1993
515'.353—dc20 93-28379

Ergreife die Feder

Das Mögliche ist ungeheuer. Die Sucht
nach Perfektion
zerstört das meiste. Was bleibt
sind Splitter
an denen sinnlos gefeilt wurde.

Friedrich Dürrenmatt

Contents

Preface

This monograph is an attempt to present the basic concepts of fully adaptive multilevel methods, including their mathematical theory, efficient algorithms, and flexible data structures. All these aspects are important to obtain successful results for practical problems. Additionally, I hope to show with this book that a unified approach combining these aspects leads to many new insights.

Multilevel adaptive methods have evolved rapidly over the last decade, and the development has reached a point where the different aspects of the discipline are finally coming together. This book is meant to be a reflection of this maturing discipline. However, the attempt to present all components of adaptive methods from functional analysis to software engineering within limited space and time has forced me to make compromises. Therefore, I have tried to simplify the material by concentrating on instructive prototype cases instead of representing the full generality of the ideas. The reader may also be warned that, despite my attempts towards unification, the theoretical foundation of multilevel methods in approximation theory requires a scientific language that is different from what is needed to discuss the benefits of object-oriented programming for these methods. Nevertheless, I hope that the selection of topics and the style of representation will be useful to anyone interested in multilevel adaptive methods.

The theory of multilevel methods has been studied systematically for almost three decades, with many papers appearing on the subject, especially in the past few years. It has therefore become difficult for one person to follow all of the different developments. In this monograph, I attempt to present a theoretical foundation of multilevel methods that fits especially well into my view of adaptive methods. In the references, I have included pointers to topics beyond the scope of this book, as they are accessible to me at the time of writing.

Many fewer publications are available on data structures for multilevel methods, and most of of them are limited to the implementation of special adaptive strategies. My approach attempts to go beyond this, because I believe that ultimately a more general, abstract treatment of the computing aspects will be as essential as a sound mathematical foundation.

Many of the ideas presented in this book were developed while I was visiting the Computational Mathematics Group of the University of Colorado at Denver in 1989. This visit, and my support in 1992, when a first version of this monograph was completed, were sponsored by grants of the Deutsche Forschungsgemeinschaft. I am grateful for helpful discussions with all my colleagues at the Technische Universität München and at the University of Colorado. For many comments that have helped me to improve the book, I am indebted to H. Bungartz, W. Dahmen, A. Kunoth, P. Leinen, P. Oswald, H. Yserentant, and C. Zenger. I want to express my special thanks to S. McCormick for his support and encouragement and a very careful review of my manuscript. Finally, I thank T. Gerstner for his help with the illustrations, the staff at SIAM for their friendly and efficient cooperation, and my wife Karin and my son Timo for their great support and patience.

Munich, March 1993 Ulrich Rüde

List of Figures

Chapter 1
Introduction

1.1. Purpose and motivation

This monograph is intended as a *practical guide* to the development of multilevel adaptive methods for the numerical solution of partial differential equations (PDEs). While many types of multilevel adaptive methods have been the subject of intensive theoretical and practical investigations, this volume focuses on the development of the so-called *fully adaptive multigrid method* (FAMe), and its basic components, the *virtual global grid* technique and the *multilevel adaptive relaxation* algorithms. These techniques generalize and extend existing alternative approaches, like the *multilevel adaptive technique* (MLAT) (see Brandt [27]) and the *fast adaptive composite grid method* (FAC) (see McCormick [57]). The FAMe is also related to the various adaptive multilevel finite element techniques, like the hierarchical basis method (see Yserentant [111] and Bank, Dupont, and Yserentant [8]) and the Bramble, Pasciak, and Xu (BPX) method (see [25]). Though our presentation focuses on the development of the FAMe, the discussion provides useful insight into the classical methods as well.

For the effective solution of PDEs, we believe that four basic components are necessary:

- good (high order) discretizations
- adaptivity
- fast (multilevel) solvers
- advanced software techniques

The discretization type should depend on the smoothness of the solution. Local smoothness is a basic characteristic of many physical problems, and any effective numerical technique must exploit this smoothness by high order discretizations. Besides higher order finite differences, this can be accomplished by high order finite elements (the so-called p-version of the finite element method), spectral methods, or by various kinds of extrapolation techniques. However, while the choice of appropriate discretizations for PDEs certainly merits discussion, this will not be further studied in this monograph because a comprehensive treatment would require a full additional volume. The interested reader is referred to the literature on high order finite element methods (see Babuška, Szabo, and Katz [5]) and extrapolation techniques (see Marchuk and Shaidurov [54] and Rüde [87]). Moreover, while interesting results have recently begun to emerge (see Pavarino [77] and Rüde [87], [90]), a unified theory for higher order multilevel methods is not yet available, so a general presentation seems premature.

In practical problems, the character of the solution varies over the domain. A solution that is smooth in certain subdomains may have singularities or boundary layers in other parts of the domain. Such structures can only be resolved by significantly reduced mesh sizes, making global uniform meshes inefficient for many practical calculations. In addition to an adaptation of the mesh to accommodate these variations, we will ultimately also need an adaptation of the approximation order. To support the adaptation, we will consider various mesh structures, including the *virtual global grid technique*, which can be understood as a meta-method and which can be used to enhance conventional grid structures.

Even if a physical problem is discretized well on adapted meshes, the resulting algebraic system will be large, so that the most efficient solvers are required. At present, some of the best general algorithms are derived from the multilevel principle. In well-designed methods, the multilevel structure is automatically built within the adaptive refinement process, so that multilevel solution algorithms are natural.

Finally, all methods, algorithms, and data structures must be implemented on a given computer, possibly of parallel type. Software design aspects have so far been neglected in the scientific discussion of numerical methods. The continuing development, however, leads to an ever increasing software complexity. This, and the spreading of languages like C++ that support modern computer science concepts without losing too much efficiency, have recently led to a renewed interest in the systematic study of programming techniques for numerical applications. This monograph attempts to present algorithmic aspects together with an analysis of data structures and software issues in a homogeneous framework. We believe that further progress with practical PDE problems will increasingly depend on developing and using appropriate advanced software design methods.

Recent advances in the theory of multilevel methods have been an additional motivation in writing this monograph. Though a multigrid convergence theory has been available since the late seventies and has continuously evolved since then, a new approach first developed by Oswald [70] provides an elegant link between multilevel methods and the theory of function spaces. The value of this approach, in our opinion, is not so much that it provides new or stronger convergence results for multilevel methods. Beyond this, it relates the finite-dimensional (multilevel) approximation spaces to the infinite-dimensional solution spaces of the PDE directly. Besides showing the optimal convergence behavior of multilevel methods, the resulting theory can be used to derive error estimates and to guide mesh refinement strategies. In particular, it provides a theoretical foundation for the two FAMe concepts, the *virtual global grid* and the *multilevel adaptive relaxation* techniques.

The classical solution of PDEs makes clear distinctions among discretization, solution, error estimation, and mesh refinement. This separation leads to suboptimal methods. Only if all components are linked together can a fully efficient method result, and only if the multilevel nature of the discretization is exploited can a mathematical structure result that leads to optimal solvers. The mesh refinement process and the multilevel structure must therefore be built appropriately. All components are interdependent, and cannot be partitioned into exchangeable modules. In numerical analysis, it is a common experience that only a unified general approach provides fully efficient algorithms.

Though the unified approach to multilevel methods has great theoretical advantages, it leads to a seemingly nonmodular software structure. For the software architecture, we must therefore develop a different type of modularity to enable the design of sufficiently general and efficient software.

This monograph is organized into four chapters. The remainder of the current chapter provides a brief introduction to our notation and the model problems.

Chapter 2 develops and summarizes a general theory of multilevel methods. The idea is to provide an abstract description of a multilevel structure by introducing the concept of *stable splittings*. Following the techniques developed by Oswald, Dahmen, and Kunoth (see the references at the beginning of Chapter 2), we link this setup to the theory of function spaces and finite element spaces. The solution space for an elliptic PDE can be understood as the limit case of an infinitely refined finite element space. The equivalence of the discrete and the continuous space can then be exploited to analyze the multilevel structure.

Chapter 3 describes how efficient multilevel algorithms can be developed from the abstract results of Chapter 2. The theory in §2.2 is formulated for scalar, two-dimensional, self-adjoint, second order elliptic PDEs, discretized by linear finite elements. However, the algorithms in Chapter 3 will be presented algebraically. This is motivated by the hope that the theory can be generalized. Besides recent theoretical results in this direction, this is also supported by numerical experiments, so that a presentation of the algorithms for a general elliptic PDE seems to be justified. The algorithmic techniques developed in Chapter 3 aim at the *fully adaptive multigrid method* although the results are also useful for conventional algorithms.

Chapter 4 is devoted to the discussion of data structures and software engineering issues related to adaptive multilevel methods. Instead of focusing on a single data structure, we provide a general framework useful for all multilevel adaptive algorithms. Besides outlining the mesh structure as an *abstract data type*, we discuss variants allowing for various degrees of adaptivity and computational efficiency. The *virtual global grids* are introduced as a meta-concept to augment classical mesh structures with a means for mesh adaptivity. In a final section, we discuss programming techniques for implementing the resulting algorithms and data types.

The three main chapters in this monograph are interdependent but are written so that they can be read separately. The reader is assumed to have a basic knowledge of classical multigrid or multilevel algorithms, at least on the level provided by Briggs [32]. For Chapter 2, a background in the theory of function spaces will be helpful. Chapter 3 relies heavily on concepts of linear algebra. To understand the discussion of Chapter 4, the reader should have some background in computer science and should have programming experience in a higher (preferably an object-oriented) programming language.

1.2. Notation

Spaces, operators, matrices, and sets are denoted by capital letters; real numbers, functions, and vectors use lower case. Below we list the symbols with a special meaning in this text.

Sets and domains

$\mathbb{N}$	set of positive integers
$\mathbb{N}_0$	set of non-negative integers
$\mathbb{R}$	set of real numbers
J	generic set of indices
Ω	a bounded open domain in $\mathbb{R}^2$, see §1.3
$\bar{\Omega}$	closure of Ω
$\mathring{\Omega}$	interior of Ω
$\Gamma = \partial\Omega$	boundary of domain Ω
Ω_h	see Definition 2.2.2
$\mathcal{T}, \mathcal{E}, \mathcal{N}$	sets of elements, edges, and nodes of a finite element partition, see §4.2.2
$S, \tilde{S}, S_k, \tilde{S}_k$	strictly active set, active set, see Definitions 3.1.5, 3.1.6, 3.4.1
RP, RT, RR, RC, RI, U_k, V_k	see §3.6
$\mathrm{Neigh}(i)$	set of neighbors of node i, see §4.2.3
$\mathrm{Conn}(i)$	set of connections of node i, see Definition 3.1.1
$M_1 \times M_2$	Cartesian product of sets
R	relation, see §4.1
$\mathfrak{P}(M)$	power set of M, see §4.1
$\emptyset$	empty set

Spaces

C^k, $C^k(\Omega)$	Space of k times differentiable functions
B^k, $B^k(\Omega)$	Besov spaces, see §2.2.2 and Definition 2.2.6
H^k, H_0^k, $H^k(\Omega)$	Sobolev spaces, see §1.3
L_2, $L_2(\Omega)$	see §1.3
V^*	dual space of V
$\mathbb{R}^d$	d-dimensional Euclidean space
$\mathcal{B}_j$	basis of V_j
$B_{j,i}$	basis function in $\mathcal{B}_j$
V, V_j	general Hilbert spaces

Scalar products and norms

$(\cdot,\cdot)_\Omega$	L_2-inner product, see §1.3
$a(\cdot,\cdot)_\Omega)$	energy inner product, see §1.3
$(\cdot,\cdot)$, $(\cdot,\cdot)_V$, $\langle\cdot,\cdot\rangle$, $a(\cdot,\cdot), a(\cdot,\cdot)_\Omega), b_j(\cdot,\cdot)$	abstract bilinear forms, see §2.1
$\Vert\cdot\Vert_V$	Hilbert space norm of space V
$\vert\cdot\vert_{B^k}$	Besov seminorm, see Definition 2.2.6
$\vert\cdot\vert_{H^k}$	Sobolev seminorm, see §1.3
$\Vert\cdot\Vert_{B^k}$	Besov norm, see Definition 2.2.6
$\Vert\cdot\Vert_{H^k}$	Sobolev norm, see §1.3
$\vert\vert\vert\cdot\vert\vert\vert$	additive Schwarz norm, see Definition 2.1.1
$\Vert\cdot\Vert$, $\Vert\cdot\Vert_\infty$, $\Vert\cdot\Vert_E$	discrete Euclidean, maximum, and energy norm, respectively

Parameters and Constants

h	mesh size
$\lambda(P)$	eigenvalues of operator P
κ	condition number, see Definition 2.1.2 and §3.2
$a_{i,j}$	elements of matrix A
c, $\bar{c}$, $\hat{c}$, $c_0, c_1, \ldots$	constants, independent of the mesh size or level, but possibly different at each occurrence.

Approximation theory

$E_j(v)$	Best approximation of v in V_j, see Definition 2.2.1
δ_h^l	lth order finite difference, see Definition 2.2.2
$\omega_2(t,f)$	lth order L_2-modulus of smoothness, see Definition 2.2.2

Operators, matrices, and vectors

I	identity operator
e_i	ith unit vector
A^T, x^T	transpose of a matrix or vector
P_{V_j}	subspace corrections, see Definition 2.1.3
P_V	additive Schwarz operator, see Definition 2.1.4
u, v, u_j, v_j	elements of space V and V_j, respectively
u^*	exact solution
A, A_k	stiffness matrix (for level k)
x, x_k	vector of unknowns (for level k)
b, b_k	right-hand side (for level k)
r, r_k, $\bar{r}$, $\bar{r}_k$	residual and scaled residual (on level k), see Definitions 2.6.1, 3.1.4, 3.3.1
I_k^l	restriction or interpolation operator from level k to l
$\hat{A}$	extended, semidefinite stiffness matrix, see §3.2
$\hat{A}_k$	hierarchically transformed stiffness matrix
diag(A)	diagonal part of A
H_k^{k-1}	hierarchical transformation, see Definition 3.6.1
$\tau_k, \bar{\tau}_k$	τ-terms, see Definition 3.6.2

1.3. Basics and model problems

The topic of this monograph is adaptive techniques for the numerical solution of elliptic partial differential equations. Our representation does not aim at the greatest possible generality, but attempts to illuminate theoretical and practical principles in simple, but typical, model situations. For example, we will develop the theory in L_2-based spaces (rather than in L_p) to avoid the additional parameter. As we only derive the theory for second order problems, we also only need the Sobolev space H^1 instead of the more general spaces H^k. For generalizations, the reader is referred to the references.

The algorithms themselves will be studied in the context of model problems. The prototypes for elliptic equations are the Dirichlet problem of *Laplace's* equation,

$$\begin{aligned} -\Delta u &= f(x,y) \quad \text{in } \Omega, \\ u &= g(x,y) \quad \text{on } \Gamma = \partial\Omega, \end{aligned} \tag{1.1}$$

and, somewhat more generally,

$$\begin{aligned} -\nabla \cdot D\nabla u + q \cdot u &= f(x,y) \quad \text{in } \Omega, \\ u &= g(x,y) \quad \text{on } \Gamma = \partial\Omega, \end{aligned} \tag{1.2}$$

where $D, q \in L_\infty(\Omega)$, $q \geq 0$, and $f \in L_2(\Omega)$. Equation (1.2) can be considered as a mathematical model of a stationary diffusion process. These will be our primary model problems to illustrate and discuss the techniques developed.

Many of the ideas easily extend to more general situations, like Neumann or mixed boundary conditions, non-self-adjoint equations, higher order equations, systems, or nonlinear equations. We further assume that $\Omega \subset \mathbb{R}^2$ is an open domain bounded by a polygon, excluding slits (that is, interior angles of 360 degrees).

The L_2-scalar product is defined by

$$(u, v)_\Omega = \iint\limits_\Omega u \cdot v \, dx, \tag{1.3}$$

and

$$\|u\|_{L_2} = \sqrt{(u, u)_\Omega} \tag{1.4}$$

defines the L_2-norm. As usual, the space L_2 is defined as the set of all functions with bounded L_2-norm:

$$L_2(\Omega) = \{f : \Omega \longrightarrow \mathbb{R} | \; \|f\|_{L_2} < \infty\}. \tag{1.5}$$

The *Sobolev seminorm* $|\cdot|_{H^1}$ is defined by

$$|u|_{H^1} = \sqrt{\left\|\frac{\partial u}{\partial x}\right\|_{L_2}^2 + \left\|\frac{\partial u}{\partial y}\right\|_{L_2}^2}, \tag{1.6}$$

and the *Sobolev norm* $\|\cdot\|_{H^1}$ is defined by

$$\|u\|_{H^1} = \sqrt{\|u\|_{L_2}^2 + |u|_{H^1}^2}. \tag{1.7}$$

$H_e^1(\Omega)$ is defined to be the (affine) subspace of the Sobolev space $H^1(\Omega)$ that enforces the *essential* boundary condition $u = g(x, y)$ on Γ for suitably smooth boundary values g.[1] As a special case, we will need $H_0^1(\Omega)$, the space of functions with homogeneous Dirichlet boundary conditions. On a bounded domain, for functions satisfying homogeneous Dirichlet boundary conditions, the seminorm $|\cdot|_{H^1}$ is equivalent to the Sobolev norm $\|\cdot\|_{H^1}$. The bilinear form corresponding to equation (1.1) is

$$a_\Omega(u, v) = (\nabla u, \nabla v)_\Omega, \tag{1.8}$$

and

$$a_\Omega(u, v) = (D \cdot \nabla u, \nabla v)_\Omega \tag{1.9}$$

corresponds to (1.2).

With these two *scalar products*, we can also introduce the *variational formulation* (also called weak formulation) of the problem: Find $u \in H_e^1(\Omega)$ such that

[1] The boundary values must satisfy $g \in H^{1/2}(\Gamma)$; see Hackbusch [44].

$$a_\Omega(u,v) = \Phi(v) \tag{1.10}$$

for all $v \in H_0^1(\Omega)$, where $\Phi(v) = (f,v)_\Omega$. Equation (1.10) is equivalent to a *minimization problem*

$$\min_{u \in H_e^1(\Omega)} a_\Omega(u,u) - (f,u)_\Omega. \tag{1.11}$$

For a numerical solution the domain Ω is partitioned by a *finite element triangulation* $(\mathcal{T}, \mathcal{N}, \mathcal{E})$, as discussed in Chapter 4. Assuming that the solution of (1.11) (or (1.10)) can be approximated by a continuous, piecewise polynomial function on the triangles, we can reduce the differential equation to a finite-dimensional problem. In the case of piecewise linear functions (so-called *linear finite elements*), we are naturally led to the *nodal basis* (see Chapter 4). For an introduction to finite element methods the reader is referred to standard textbooks like Axelsson and Barker [2], Braess [21], Hackbusch [44], and Strang and Fix [102].

The discretization of problems like (1.1) or (1.2) with finite elements and the representation of the solution in a nodal basis leads to a linear system

$$Ax = b, \tag{1.12}$$

where A is called the *stiffness matrix* (see Chapters 3 and 4). This is the starting point for the development of the algorithms in Chapter 3. Generally, A is *sparse*, that is, only a small fraction of the entries are nonzero. Any efficient solution technique for (1.12) must exploit the sparsity in the data structures and algorithms. Direct solution techniques based on elimination necessarily destroy the sparsity, due to *fill-in*.

This motivates the general interest in *iterative solution techniques*. For our model cases, the matrix A will be symmetric positive definite. Therefore, classical relaxation methods, like Gauß–Seidel relaxation, are directly applicable. These methods, as well as other iterative techniques, like conjugate gradient iteration, will slow down when the number of unknowns grows. This is related to the conditioning of A that gets worse when the domain is resolved with finer meshes and more degrees of freedom. To solve large problems efficiently, we must therefore consider multigrid acceleration or preconditioning techniques. Chapters 2 and 3 will present a detailed discussion of these issues. Furthermore, the development of efficient algorithms cannot be separated from the design of data structures for A, x, and b. This is discussed in Chapter 4.

Chapter 2

Multilevel Splittings

In this chapter, we collect results for the multilevel splitting of finite element spaces. Our presentation is motivated by recent results of Oswald [70], [75], [76] and Dahmen and Kunoth [34]. These papers are in turn related to the quickly developing theory of multilevel preconditioners as studied by Yserentant [111], [112], [113], [114], Xu [108], [109], Bramble, Pasciak, and Xu [25], Dryja and Widlund [37], [38], S. Zhang [115], and X. Zhang [116]. The approach by Oswald, Dahmen, and Kunoth is based on results from the theory of function spaces. The relationship of this abstract theory to multilevel methods is developed in a sequence of papers and reports [34], [69], [68], [70], [71], [72], [74], [73], [75], [76].

As the basic algorithmic structure, we introduce the so-called *multilevel additive Schwarz* method. The idea is to use a hierarchy of levels for a multiscale representation of the problem and to combine the contributions of all levels in a sum. This process implicitly defines an *operator sum* that is well behaved and that has bounded condition number independent of the number of levels. Thus, it is suitable for fast iterative inversion by conjugate gradient-type algorithms.

The recent theoretical approach to these methods by Oswald (see the papers cited above) is based on results in approximation theory, in particular on methods from the theory of Besov spaces. The relevant basic results can be found in Nikol'skiĭ [66] and Triebel [105]. An outline of these results, including a bibliography, is also given in a survey article by Besov, Kudrayavtsev, Lizorkin, and Nikol'skiĭ [15].

From a more general perspective, the multilevel additive Schwarz method is also related to multigrid methods and their theory. Classical multigrid methods can be interpreted as a *multiplicative Schwarz* method where the levels are visited sequentially and the basic structure is a *product of operators*.

Multigrid convergence theory has been studied in so many papers that a complete bibliography cannot be given in the context of this monograph. We refer to Hackbusch [43] and McCormick [56] for the classical theory and to Yserentant [114] for a review of recent developments.

It should be noted that the interpretation of multilevel techniques as

Schwarz methods uses the structure of nested spaces and symmetric operators corresponding to what is known as *variational multigrid.* The classical multigrid theory is more general in this respect, because it assumes relations only between the (possibly nonsymmetric) operators; it assumes no special relations between the grid spaces.

The unified multigrid convergence theory developed from the Schwarz concept seems to need nesting of the spaces and symmetry of the operators (however, see attempts to generalize these limitations by Bramble, Pasciak, and Xu [26] and Xu [110]). At this stage, the new theory also fails to describe some features of the multigrid principle, like the dependency of the performance on the number of smoothing steps per level. Typically, however, the new theory does not need as strong regularity assumptions as the classical multigrid theory. The interested reader is referred to the original papers by Dryja and Widlund [38], Bramble, Pasciak, Wang, and Xu [24], [23], [25], [109], and Zhang [116], where the theory is developed.

Here, we will describe these new techniques, following closely the approach by Oswald, Dahmen, and Kunoth, because they provide an elegant theoretical foundation of the fast adaptive methods that will be discussed in Chapter 3. Our emphasis here is to give a consistent presentation of the abstract foundation in approximation theory and its application to the prototype finite element situations that arise in the solution of the model problems in §1.3. In particular, we will show that the same theoretical background can be used to justify fast iterative solvers, error estimates, and mesh refinement strategies.

2.1. Abstract stable splittings

The basis of multilevel algorithms is a decomposition or splitting of the solution space into subspaces. Multilevel algorithms depend on this structure and its particular features. To classify multilevel splittings, we introduce the notion of a *stable splitting* that we will describe in an abstract setting.

We assume that the basic space V is a Hilbert space equipped with a scalar product $\langle\cdot,\cdot\rangle_V$ and the associated norm

$$\|u\|_V = \langle u, u\rangle_V^{1/2}.$$

The elliptic partial differential equation is formulated with a V-elliptic, symmetric, continuous bilinear form $a : V \times V \longrightarrow \mathbb{R}$. Thus there exist constants $0 < c_1 \le c_2 < \infty$ such that

$$c_1\langle v, v\rangle_V \le a(v,v) \le c_2\langle v, v\rangle_V \tag{2.1}$$

for all $v \in V$. In view of the model problems (1.1)–(1.2) and their variational form, we study the abstract problem: Find $u \in V$ such that

$$a(u,v) = \Phi(v) \tag{2.2}$$

for all $v \in V$, where the functional $\Phi \in V^*$ is a continuous linear form.

To introduce a multilevel structure we consider a finite or infinite collection $\{V_j\}_{j\in J}$ of subspaces of V, each with its own scalar product $(\cdot,\cdot)_{V_j}$ and the associated norm

$$\|u\|_{V_j} = (u,u)_{V_j}^{1/2}.$$

We further assume that the full space V can be represented as the sum of the subspaces V_j, $j \in J$,

$$V = \sum_{j\in J} V_j. \tag{2.3}$$

REMARK 2.1.1. *Later we will additionally assume that the spaces are nested, that is,* $J \subset \mathbb{N}_0$, $V_i \subset V_j$, *if* $i \le j$. *The theory in this section, however, does not depend on this assumption and without it many more iterative methods can be described in the abstract framework, including classical relaxation methods, block relaxation, and domain decomposition.*

REMARK 2.1.2. *In typical applications,* $\|\cdot\|_V$ *is equivalent to the* H^1*-Sobolev norm. It is our goal to build equivalent norms based on the subspaces* V_j *and their associated norms. The subspace norms* $\|\cdot\|_{V_j}$ *will be properly scaled* L_2*-norms. Based on the associated bilinear forms, we can construct a multilevel operator that uses elementary operations (based on the* L_2*-inner products) on all levels, except possibly the coarsest one; see Definitions* 2.1.3 *and* 2.1.4. *If this operator is spectrally equivalent to the original operator, it can be used to build efficient preconditioners, error estimates, and refinement algorithms.*

The system of subspaces induces a structure in the full space. Any element of V can be represented as a sum of elements in V_j, $j \in J$. Generally, this representation is nonunique. This observation gives rise to the following definition.

DEFINITION 2.1.1. *The* **additive Schwarz norm** $|||\cdot|||$ *in* V *with respect to the collection of subspaces* $\{V_j\}_{j\in J}$ *is defined by*

$$|||v||| \stackrel{\text{def}}{=} \inf\left\{ \left(\sum_{j\in J} \|v_j\|_{V_j}^2\right)^{1/2} \,\middle|\, v_j \in V_j,\ \sum_{j\in J} v_j = v \right\}. \tag{2.4}$$

As we will show, how well a multilevel algorithm converges depends on how well the multilevel structure captures the features of the original problem, that is, how well the additive Schwarz norm approximates the original norm $\langle\cdot,\cdot\rangle_V$ in V. This motivates the following definition.

DEFINITION 2.1.2. *A collection of spaces* $\{V_j\}_{j\in J}$ *is called a* **stable splitting** *of* V *if*

$$\sum_{j\in J} V_j = V$$

and if $\|\cdot\|_V$ *is equivalent to the additive Schwarz norm of* V, *that is, if there exist constants* $0 < c_3 \le c_4 < \infty$ *such that*

$$c_3\|v\|_V^2 \le |||v|||^2 \le c_4\|v\|_V^2 \tag{2.5}$$

for all $v \in V$. The number

$$\kappa(V, \{V_j\}_{j\in J}) \stackrel{\text{def}}{=} \inf(c_4/c_3), \tag{2.6}$$

that is, the infimum over all possible constants in (2.5), *is called the* **stability constant** *of the splitting* $\{V_j\}_{j\in J}$.

REMARK 2.1.3. *If V is finite-dimensional, any splitting is stable. The concept of a stable splitting is therefore primarily relevant in an infinite-dimensional setting. In practice, the problem in a finite-dimensional space is not to show the existence of a stable splitting, but to study the size of the stability constant. We will assume that the finite-dimensional discrete space is embedded in an infinite-dimensional space. By showing that the splitting of the infinite-dimensional space is stable, we can derive bounds for the stability constant that are uniform in the number of levels.*

The definition of a stable splitting leaves room for many cases, including pathological ones.

Example. Consider a splitting of an arbitrary nontrivial Hilbert space V into two subspaces V_1 and V_2. Let $V_1 = \text{span}\{x\}$ for some $x \in V$, $x \neq 0$, and

$$\|\alpha x\|_{V_1} \stackrel{\text{def}}{=} |\alpha|,$$

where $\alpha \in \mathbb{R}$. Then let $V_2 = V$ and $\|\cdot\|_{V_2} \equiv \|\cdot\|_V$. To show that the splitting $V = V_1 + V_2$ is stable, we must show that (2.5) holds. In this simple case we have

$$|||v||| = \inf_{v_1\in V_1} \sqrt{\|v_1\|_{V_1}^2 + \|v - v_1\|_{V_2}^2} = \inf_{\alpha\in\mathbb{R}} \sqrt{\alpha^2 + \|v - \alpha x\|_V^2}.$$

Therefore, the upper bound holds trivially with $c_4 = 1$ (set $\alpha = 0$). The lower bound can be constructed as follows:

$$\begin{aligned}
|||v|||^2 &= \inf_{\alpha\in\mathbb{R}} \left(\alpha^2 + \|v - \alpha x\|_V^2\right) \\
&\geq \inf_{\alpha\in\mathbb{R}} \left(\alpha^2 + (\|v\|_V - \alpha\|x\|_V)^2\right) \\
&= \inf_{\alpha\in\mathbb{R}} \left((1 + \|x\|_V^2)\alpha^2 - 2\|x\|_V\|v\|_V\alpha + \|v\|^2\right) \\
&= \frac{-(\|x\|_V\|v\|_V)^2 + \|v\|_V^2(1 + \|x\|_V^2)}{1 + \|x\|_V^2} \\
&= c\|v\|_V^2,
\end{aligned}$$

where $c = 1/(1 + \|x\|_V^2)$. Despite the stability, the practical value of the splitting is doubtful. As we will see further below, $V_2 = V$ and $\|\cdot\|_{V_2} = \|\cdot\|_V$ imply that methods based on this splitting involve the solution of a problem that is equivalent to the original one.

The following example is more typical for the situations that are of interest to us.

Example. Let the one-dimensional *Fourier components* ϕ_n be given by

$$\phi_n : \begin{cases} [0,1] & \longrightarrow & \mathbb{R}, \\ \phi_n(x) & = & \sin(n\pi x), \end{cases}$$

$n \in \mathbb{N}$. Consider the space

$$V = \left\{ f = \sum_{n=1}^{\infty} a_n \phi_n \mid \|f\|_V < \infty \right\}, \tag{2.7}$$

where $a_n \in \mathbb{R}$ and

$$\left\| \sum_{n=1}^{\infty} a_n \phi_n \right\|_V \stackrel{\text{def}}{=} \sqrt{\sum_{n=1}^{\infty} n^2 a_n^2}. \tag{2.8}$$

Subspaces of V are now defined by

$$V_j = \left\{ f = \sum_{n=1}^{2^j-1} a_n \phi_n \right\} \subset V, \tag{2.9}$$

where $j \in \mathbb{N}$ and the corresponding norms are defined by

$$\left\| \sum_{n=1}^{2^j-1} a_n \phi_n \right\|_{V_j} \stackrel{\text{def}}{=} 2^j \sqrt{\sum_{n=1}^{2^j-1} a_n^2}. \tag{2.10}$$

Note that $\|\cdot\|_V$ corresponds to an H^1-norm, while $\|\cdot\|_{V_j}$ corresponds to a (scaled) L_2-norm.

Next, we show that the spaces $(V_j, \|\cdot\|_{V_j})$, $j \in \mathbb{N}$, are a stable splitting of $(V, \|\cdot\|_V)$. Clearly, each $f \in V$ can be decomposed in the form

$$f = \sum_{n=1}^{\infty} a_n \phi_n = \sum_{j=1}^{\infty} \underbrace{\sum_{n=1}^{2^j-1} a_{nj} \phi_n}_{= f_j \in V_j}, \tag{2.11}$$

where $\sum_{j=1}^{\infty} a_{nj} = a_n$. Setting $a_{nj} = 0$ if $n \geq 2^j$ we have

$$\sum_{j=1}^{\infty} \|f_j\|_{V_j}^2 = \sum_{j=1}^{\infty} 2^{2j} \sum_{n=1}^{2^j-1} a_{nj}^2 = \sum_{n=1}^{\infty} \sum_{j=1}^{\infty} 2^{2j} a_{nj}^2,$$

so that

$$|||f||| = \inf_{f_j \in V_j, \sum f_j = f} \|f_j\|_{V_j}^2 = \inf_{\sum a_{nj} = a_n} \sum_{n=1}^{\infty} \sum_{j=1}^{\infty} 2^{2j} a_{nj}^2 = \sum_{n=1}^{\infty} 2^{2j_n} a_n^2, \tag{2.12}$$

where j_n is the smallest positive integer such that $2^{j_n} > n$. Clearly,

$$n \leq 2^{j_n} \leq 2n,$$

so that the lower bound in (2.5) holds with $c_3 = 1$ and the upper bound holds with $c_4 = 2^2$.

We now continue our discussion of the abstract theory of stable splittings. To describe multilevel algorithms we further introduce V_j-elliptic, symmetric, bilinear forms

$$b_j : V_j \times V_j \longrightarrow \mathbb{R}$$

in the spaces V_j, respectively. These bilinear forms give us the flexibility to describe a wider class of multilevel algorithms within our framework. The particular choice of b_j will lead to different multilevel algorithms. In a first reading one may think of $b_j(\cdot,\cdot) \equiv (\cdot,\cdot)_{V_j}$. Generally, for a properly working multilevel algorithm, we require that the b_j are equivalent to the respective inner product of the subspace, that is, that there exist constants $0 < c_5 \leq c_6 < \infty$ such that

$$c_5(v_j, v_j)_{V_j} \leq b_j(v_j, v_j) \leq c_6(v_j, v_j)_{V_j}, \tag{2.13}$$

for all $v_j \in V_j$, $j \in J$.

Based on these V_j-elliptic bilinear forms, the components of a multilevel algorithm can be defined. In our setup, multilevel algorithms are described in terms of *subspace corrections*, mapping the full space V into each of the subspaces V_j.

DEFINITION 2.1.3. *The mappings* $P_{V_j} : V \longrightarrow V_j$ *are defined by the variational problem*

$$b_j(P_{V_j}u, v_j) = a(u, v_j), \tag{2.14}$$

for all $v_j \in V_j$, $j \in J$. *Analogously, we define* $\phi_j \in V_j$ *by*

$$b_j(\phi_j, v_j) = \Phi(v_j), \tag{2.15}$$

for all $v_j \in V_j$, $j \in J$.

With the subspace corrections we can define the *additive Schwarz operator* as follows.

DEFINITION 2.1.4. *The* **additive Schwarz operator** *(also called BPX operator)* $P_V : V \longrightarrow V$ *with respect to the multilevel structure on* V *(that is,* $a(\cdot,\cdot)$, $\{V_j\}_{j\in J}$, *and* $b_j(\cdot,\cdot)$*) is defined by*

$$P_V = \sum_{j\in J} P_{V_j}. \tag{2.16}$$

Analogously, $\phi \in V$ *is defined by*

$$\phi = \sum_{j\in J} \phi_j. \tag{2.17}$$

REMARK 2.1.4. *The operator* P_V *provides the basis for the so-called* additive Schwarz method. *In Theorem* 2.1.1 *below we will also show that* P_V *can be used to build problems equivalent to the discrete variational problem (see equation* (2.18)*), but with much better conditioning, so that they can be solved efficiently by iterative techniques.*

REMARK 2.1.5. *With a suitably defined bilinear form b_j, it is possible to evaluate P_V efficiently based on its definition as a sum. The explicit construction of P_V, which would be inefficient, is not required.*

REMARK 2.1.6. *Many iterative algorithms, including Jacobi iteration, block-Jacobi, domain decomposition, and variational multigrid methods, can be described in the abstract framework given in Definitions* 2.1.1–2.1.4. *Most of these methods, however, do not generate a stable splitting of the infinite-dimensional function space. The hierarchical structure in the subspace system seems to be essential for obtaining a stable splitting. Otherwise, the complexity of the original problem would have to be captured in the bilinear forms $b_j(\cdot,\cdot)$, and then the evaluation of the P_{V_j} would be as expensive as the solution of the original problem itself (see also the examples above).*

We conclude this abstract discussion by stating and proving two theorems that show the relationship between the concept of a stable splitting and the properties of the additive Schwarz operator. Based on Definitions 2.1.1–2.1.4, the following theorem holds.

THEOREM 2.1.1. *Assume that the subspaces V_j, $j \in J$, of a Hilbert space V are a stable splitting. Assume further that P_{V_j} and P_V are defined as in Definitions* 2.1.3 *and* 2.1.4 *with bilinear forms b_j satisfying* (2.13). *The variational problem* (2.2) *is equivalent to the operator equation*

$$P_V u = \phi, \tag{2.18}$$

and the spectrum of P_V can be estimated by

$$\frac{c_1}{c_4 c_6} \le \lambda_{\min}(P_V) \le \lambda_{\max}(P_V) \le \frac{c_2}{c_3 c_5}. \tag{2.19}$$

Proof. The unique solvability of (2.18) is a consequence of the positive definiteness asserted in (2.19) that we will prove below. The solution of (2.18) coincides with the solution of (2.2) by the definition of P_V and ϕ; see Definition 2.1.4 and equations (2.16), (2.17), respectively.

We now establish the lower bound of the spectrum asserted in (2.19). Let $u_j \in V_j$, $j \in J$, be an arbitrary decomposition of $u \in V$, that is, $\sum_{j\in J} u_j = u$. Then

$$\begin{aligned} a(u,u) &= \sum_{j\in J} a(u_j,u) = \sum_{j\in J} b_j(P_{V_j}u, u_j) \\ &\le \left(\sum_{j\in J} b_j(P_{V_j}u, P_{V_j}u)\right)^{1/2} \left(\sum_{j\in J} b_j(u_j,u_j)\right)^{1/2} \\ &\le \left(\sum_{j\in J} a(P_{V_j}u, u)\right)^{1/2} \left(\sum_{j\in J} c_6 (u_j,u_j)_{V_j}\right)^{1/2}. \end{aligned}$$

Taking the infimum of all decompositions of the form

$$\sum_{j\in J} u_j = u$$

we get

$$\begin{aligned} a(u,u) &\leq a(P_V u, u)^{1/2} c_6^{1/2} |||u||| \\ &\leq a(P_V u, u)^{1/2} (c_6 c_4)^{1/2} \|u\|_V \\ &\leq a(P_V u, u)^{1/2} (c_6 c_4 / c_1)^{1/2} a(u,u)^{1/2}. \end{aligned}$$

Therefore,

$$a(P_V u, u) \geq \frac{c_1}{c_4 c_6} a(u,u). \tag{2.20}$$

This establishes the lower bound in (2.19). Thus P_V is invertible so that we can define a uniquely determined $z \in V$ that satisfies $P_V z = u$. Hence,

$$\begin{aligned} a(P_V^{-1} u, u) = a(z, P_V z) &= \sum_{j \in J} a(z, P_{V_j} z) \\ &= \sum_{j \in J} b_j(P_{V_j} z, P_{V_j} z) \\ &\geq c_5 \sum_{j \in J} (P_{V_j} z, P_{V_j} z)_{V_j} \\ &\geq c_5 |||P_V z|||^2 \\ &\geq c_3 c_5 \|P_V z\|_V^2 \\ &\geq \frac{c_3 c_5}{c_2} a(u,u). \end{aligned}$$

We conclude that

$$a(v, P_V v) \leq \frac{c_2}{c_3 c_5} a(v,v)$$

for all $v \in V$. This yields the upper bound on the spectrum asserted in (2.19).

REMARK 2.1.7. *Theorem* 2.1.1 *shows that the additive Schwarz method generates a well-conditioned operator if the splitting of the space is stable.*

REMARK 2.1.8. *Results related to Theorem* 2.1.1 *have been shown by other authors, in many cases restricted to special cases like* $V = H^1(\Omega)$. *The interested reader is referred to, e.g., Yserentant* [111], *Bramble, Pasciak, and Xu* [25], *Dryja and Widlund* [38], *and Zhang* [116]. *Our presentation of Theorem* 2.1.1 *has followed Oswald* [75].

Computationally, applying the additive Schwarz operator P_V amounts to transferring the residual to all subspaces V_j, applying the inverse of the operator defined by b_j in each subspace, and finally collecting the interpolated results back in V. In the language of multigrid methods, the transfer to V_j is a *restriction.* The b_j implicitly define which kind of smoothing process is used. In the simplest case, the restricted residual is only scaled appropriately, corresponding to a *Richardson smoother.*

From the perspective of classical multigrid, it may be surprising that the *additive* operator has a uniformly bounded condition number, meaning that effective solvers with multigrid efficiency can be obtained by applying steepest

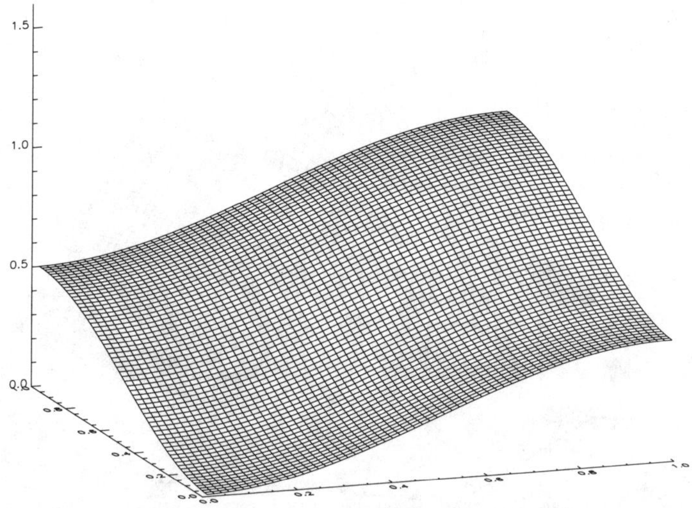

FIG. 2.1. *Idealized spectrum of the discrete Laplace operator.*

descent or conjugate gradient iteration to this system. Classical multigrid is formulated as a *sequential* algorithm where the levels are visited one after the other. The multilevel additive Schwarz method is the corresponding *simultaneous* variant, where the work on the levels is done in parallel (however, see Remark 2.1.9). The sequential treatment of the levels is not necessary for obtaining optimal orders of efficiency.

To illustrate the method, we now visualize the effect of the additive Schwarz process. We assume that V describes the solution space of the discretized two-dimensional Laplacian. In Fig. 2.1 we visualize the spectrum of the discretized Laplace operator. The figure shows the eigenvalues, where the x- and y-axes in the plot represent the frequencies with respect to the two original spatial directions. Each eigenvalue is represented by a point on the graph of the function

$$f(x,y) = \tfrac{1}{2} - (\cos(\pi x) + \cos(\pi y))/4 \qquad (2.21)$$

for $h \leq x, y \leq 1$, where h is the mesh size. The x- and y-coordinates specify the frequency of the Fourier mode relative to the mesh size. Here, we have scaled the discrete Laplacian such that the extreme eigenvalues occur in the northeast corner (near $(1,1)$) and the southwest corner of the frequency domain, where the values are $O(1)$ and $O(h^2)$, respectively. Thus, for this example problem, the condition number grows with $O(1/h^2)$. Consequently, iterative methods, like steepest descent, need a number of cycles to converge, that is, proportional to $O(h^{-2})$.

The additive Schwarz method additionally transfers the residual to coarser levels with mesh sizes $2h, 4h, 8h, \ldots$. This transfer is idealized by restricting the full spectrum represented by $(h,1) \times (h,1)$ to the squares $(h,0.5) \times (h,0.5)$,

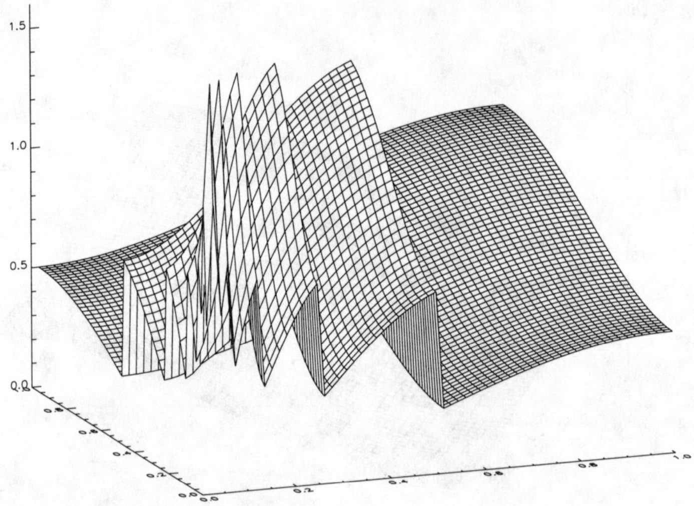

FIG. 2.2. *Idealized spectrum of the additive Schwarz operator associated with the Laplacian.*

$(h, 0.25) \times (h, 0.25), \ldots$. On the coarser levels, the result is rescaled, such that the maximum value in the restricted spectrum is again 1. The results of all levels are finally extended to the full spectrum and are added together. The result of this process is displayed in the function plot of Fig. 2.2.

The multilevel sum of operators, whose eigenvalues are represented in Fig. 2.2, seems to have a minimal eigenvalue bounded away from 0 and a maximal eigenvalue not much larger than the Laplacian itself. The plot suggests that the minimum and maximum value of the combined spectrum are bounded independently of the number of levels.

The method, as it is discussed in this idealized setting, is impractical because the transfer between levels by exact cut-off functions in Fourier space cannot be implemented efficiently. The usual transfer operations are only approximate cut-off functions in Fourier space. This would have to be taken into account in a rigorous analysis.

REMARK 2.1.9. *For computing the additive Schwarz operator P_V applied to a function one may think of evaluating the terms $P_{V_j}u$ in the sum in parallel. This can be exploited most efficiently if the spaces are non-nested, e.g., when $\{V_j\}_{j \in J}$ arise from a domain decomposition. Classical domain decomposition with many subdomains and without coarse mesh spaces, however, does* **not** *cause stable splittings. These usually depend in an essential way on a hierarchical structure, often a nesting of the spaces. Unfortunately, a straightforward parallelization of the sum for nested spaces is often inefficient, since the terms in the sum naturally depend on each other. More precisely, in the case of global meshes, the usual way to compute $P_{V_i}u$ automatically* *generates $P_{V_j}u$ for all $j > i$. Thus, an optimized implementation treating the levels sequentially may be equally fast, but will need only one processor. The multilevel additive Schwarz method must therefore be parallelized using techniques such as those used in the parallelization of multigrid methods (see McBryan et al. [55]). These approaches are usually based on a domain decomposition, and a common problem, then, is that processors tend to go idle on coarse grids, leading to reduced parallel efficiency.*

This argument assumes that we use simple residual corrections *on each level and that the hierarchical structure is induced by* global *mesh refinement. Treating levels in parallel may be attractive, if the process on each level (the smoothing) is computationally significantly more expensive than the restriction operators, or when the mesh structure is highly nonuniform. One such case is the* asynchronous fast adaptive composite grid *(AFAC) method, introduced by McCormick* [57].

To illustrate the relationship between the additive Schwarz norm of Definition 2.1.1 and the additive Schwarz operator of Definition 2.1.4, we now study the special case, when the bilinear forms $a(\cdot,\cdot)$ in V and $b_j(\cdot,\cdot)$ in V_j coincide with the natural bilinear forms on the respective spaces. This is analyzed in the following theorem.

THEOREM 2.1.2. *If*

$$b_j(\cdot,\cdot) \equiv (\cdot,\cdot)_{V_j} \tag{2.22}$$

for all $j \in J$ and

$$a(\cdot,\cdot) \equiv \langle\cdot,\cdot\rangle_V, \tag{2.23}$$

then

$$|||u|||^2 = \langle P_V^{-1}u, u\rangle_V \tag{2.24}$$

for all $u \in V$.

Proof. With (2.22) and (2.23) the definition of P_{V_j} reads

$$(P_{V_j}u, v_j)_{V_j} = \langle u, v_j\rangle_V \tag{2.25}$$

for all $u \in V$, $v_j \in V_j$, and $j \in J$. Let z be defined by $P_V z = u$. This is possible because P_V is positive definite according to Theorem 2.1.1. We have

$$\begin{aligned}
\langle P_V^{-1}u, u\rangle_V &= \langle z, P_V z\rangle_V \\
&= \sum_{j\in J}\langle z, P_{V_j}z\rangle_V \\
&= \sum_{j\in J}(P_{V_j}z, P_{V_j}z)_{V_j} \\
&\geq |||u|||^2 = \inf\left\{\sum_{j\in J}\|u_j\|_{V_j}^2 \,\middle|\, u_j \in V_j, \sum_{j\in J} u_j = u\right\}
\end{aligned}$$

because

$$u = P_V z = \sum_{j\in J} P_{V_j} z.$$

Finally, we show that this particular splitting attains the infimum. We do this by choosing an arbitrary splitting $v_j \in V_j$, $j \in J$, such that $\sum_{j\in J} v_j = u$, and showing that it yields a larger or equal sum of norms.

$$
\begin{aligned}
\sum_{j\in J}(v_j, v_j)_{V_j} &= \sum_{j\in J}\Big((P_{V_j}z, P_{V_j}z)_{V_j} + 2(v_j - P_{V_j}z, P_{V_j}z)_{V_j} \\
&\qquad + (v_j - P_{V_j}z, v_j - P_{V_j}z)_{V_j}\Big) \\
&= \sum_{j\in J}\Big((P_{V_j}z, P_{V_j}z)_{V_j} + 2\langle v_j - P_{V_j}z, z\rangle_V \\
&\qquad + (v_j - P_{V_j}z, v_j - P_{V_j}z)_{V_j}\Big) \\
&= 2\langle \underbrace{(u - P_V z), z\rangle_V}_{=0} \\
&\qquad + \sum_{j\in J}\left((P_{V_j}z, P_{V_j}z)_{V_j} + \underbrace{(v_j - P_{V_j}z, v_j - P_{V_j}z)_{V_j}}_{\geq 0}\right) \\
&\geq \sum_{j\in J}(P_{V_j}z, P_{V_j}z)_{V_j}.
\end{aligned}
$$

This concludes the proof.

REMARK 2.1.10. *If the bilinear forms that define P_V coincide with the natural ones in V_j and V, respectively, then Theorem 2.1.2 shows that P_V^{-1} defines the bilinear form associated with the norm $|||\cdot|||$.*

2.2. Finite element approximation spaces in two dimensions

In this section we will apply the concept of stable splittings in the context of finite element spaces. Typically, the full space $V = H^1(\Omega)$ will correspond to the function space associated with the partial differential equation, and the $\{V_j\}_{j\in\mathbb{N}_0}$ will be an infinite collection of subspaces generated by successively refining a finite element approximation of the differential equation. To apply the results of §2.1 we will consider splittings of V in a *nested* sequence of spaces

$$
V_0 \subset V_1 \subset V_2 \subset \cdots \subset V = H_0^1(\Omega) \tag{2.26}
$$

generated by a regular family of nested triangulations

$$
\mathcal{T}_0 \subset \mathcal{T}_1 \subset \mathcal{T}_2 \subset \cdots
$$

of a bounded domain $\Omega \subset \mathbb{R}^2$. The proper structure for such finite element partitions is discussed in Chapter 4. For a detailed presentation of the properties required for finite element partitions, see also Ciarlet [33]. We assume that continuous, piecewise linear elements are used, corresponding to the second order model problems of §1.3. For generalizations to more complicated situations, see the references listed at the beginning of this chapter.

For the sake of simplicity we will also assume that the domain Ω satisfies the *uniform cone condition.* The main consequence of this restriction in our context is that the slit domain with an interior angle of 360 degrees is excluded. Our results can be extended even to such domains by additional considerations.

We assume that V_j is equipped with the inner product

$$(u_j, v_j)_{V_j} = 2^{2j}(u_j, v_j)_{L_2} \tag{2.27}$$

for $u_j, v_j \in V_j$, and $j \in \mathbb{N}$. The associated norm is defined as

$$\|\cdot\|_{V_j} = (\cdot,\cdot)_{V_j}^{1/2}.$$

For the coarsest space, we use

$$(u_0, v_0)_{V_0} = \langle u_0, v_0\rangle_V, \tag{2.28}$$

for $u_0, v_0 \in V_0$, that is, the inner product (and H^1-norm) inherited from the full space.

REMARK 2.2.1. *The* coarsest space V_0 *plays an exceptional role and will have an inner product inducing a norm that is equivalent to the* H^1*-norm, inherited from* V. *If* $\|\cdot\|_{V_0}$ *is induced by the full space norm* $\|\cdot\|_V$ *(algorithmically, this corresponds to solving the coarsest level equations exactly), then, roughly speaking, the multilevel technique is independent of the initial mesh and the associated space. If the coarsest level is only equipped with an* L_2*-equivalent bilinear form (meaning that on the coarsest level only a preconditioner is used), then it is essential that we demand that the coarsest triangulation* $\mathcal{T}_0$ *not be too fine. Typically, we may require that the characteristic mesh size* h_0 *(see* §4.2.4*) of the coarsest triangulation is of the same order as the size of the domain, that is,* $h_0 = O(\text{diam}(\Omega))$. *Then, on* V_0, *the* H^1*-norm and* L_2*-norm (or any other norm) will be equivalent. A careful analysis of the influence of the coarsest level triangulation is given in Bornemann and Yserentant* [20].

2.2.1. Approximation property, inverse property, and stable bases.

To further develop this concept we need several definitions from approximation theory.

DEFINITION 2.2.1. *The* **best approximation** *of* $v \in V$ *in* $V_j \subset V$ *is defined by*

$$E_j(v) = \inf\{\|v - v_j\|_{L_2} |\ v_j \in V_j\}. \tag{2.29}$$

For $a, b \in \mathbb{R}^2$ we use

$$[a, b] = \{\lambda a + (1-\lambda)b \mid 0 \le \lambda \le 1\}$$

to denote the line segment from a to b. Furthermore, for $a = (a_1, a_2) \in \mathbb{R}^2$,

$$|a| \stackrel{\text{def}}{=} \sqrt{a_1^2 + a_2^2}$$

denotes the *Euclidean norm* of a. The moduli of smoothness are now introduced in the following definitions.

DEFINITION 2.2.2. *For each* $h \in \mathbb{R}^2$, Ω_h *is defined as the subset of* Ω *such that the whole line segment* $[x, x+h]$ *is contained in* Ω:

$$\Omega_h = \{x \in \Omega |\ [x, x+h] \subset \Omega\};$$

δ_h^l *is defined as the* l**th order finite difference** *with step size* h:

$$\delta_h^l f(x) = \sum_{i=0}^{l} \binom{l}{i} (-1)^{l-i} f(x+ih);$$

and $\omega_l(t, f)$ *is the* l**th order** L_2**-modulus of smoothness**, *defined by*

$$\omega_l(t, f) = \sup_{|h| \le t} \{\|\delta_h^l f\|_{L_2(\Omega_{lh})}\}.$$

With the best approximation and the moduli of smoothness we can define an important feature of a multilevel structure.

DEFINITION 2.2.3. *A system of subspaces* $V_j \subset V$, $j \in \mathbb{N}_0$, *satisfies the* **approximation property** *if there exists a constant* $0 < c < \infty$ *such that*

$$E_j(u) \le c\omega_2(2^{-j}, u) \tag{2.30}$$

for all u *in* $H^1(\Omega)$.

REMARK 2.2.2. *Equation* (2.30) *is a* Jackson-type *inequality; see Oswald* [76] *for more general definitions. The approximation property* (2.30) *is related to approximation estimates of the form*

$$E_j(v) \le c2^{-jm}|v|_{H^m(\Omega)} \tag{2.31}$$

for $j > 0$ *and for all* $v \in H^m(\Omega)$, *by the so-called* K-functionals; *see Dahmen and Kunoth* [34, *Prop.* 4.1.] *and Johnen and Scherer* [48].

Next, we define an *inverse property* using the L_2-moduli of smoothness.

DEFINITION 2.2.4. *The collection of subspaces* $V_j \subset V$, $j \in \mathbb{N}_0$, *satisfies the* **inverse property** *if there exists a constant* $0 < c < \infty$ *and a real number* $\lambda > 1$ *such that*

$$\omega_2(t, v_j) \le c2^{\lambda j} t^{\lambda} \|v_j\|_{L_2} \tag{2.32}$$

for $0 < t \le 2^{-j}$ *and arbitrary* $v_j \in V_j$, $j \in \mathbb{N}_0$.

REMARK 2.2.3. *For linear finite elements, equation* (2.32) *is a* Bernstein-type *inequality* (*see Oswald* [76] *and Dahmen and Kunoth* [34] *for the definition in a more general setting*).

REMARK 2.2.4. *Equation* (2.32) *implies that*

$$\|v_j\|_{H^1} \le c2^j \cdot \|v_j\|_{L_2} \tag{2.33}$$

for all $v_j \in V_j$, $j \in \mathbb{N}_0$.

REMARK 2.2.5. *Equation* (2.32) *holds with* $\lambda = 3/2$. *This is shown in Oswald* [70, *Lemma* 3].

DEFINITION 2.2.5. *A collection of bases* $\mathcal{B}_j$ *for the respective spaces in* V_j, $j \in \mathbb{N}_0$, *where*

$$\mathcal{B}_j = \{B_{j,i} |\; i = 1, 2, \ldots, n_j;\; n_j = \dim(V_j)\},$$

such that $V_j = \operatorname{span}(\mathcal{B}_j)$, *is called* **stable** *if there exist constants* $0 < c_1 \le c_2 < \infty$ *independent of* $j \in \mathbb{N}_0$ *and constants* d_j *such that*

$$c_1 d_j \sum_i^{n_j} \beta_{j,i}^2 \le \|v_j\|_{L_2}^2 \le c_2 d_j \sum_i^{n_j} \beta_{j,i}^2 \tag{2.34}$$

for all $v_j \in V_j$, *where the coefficients* $\beta_{j,i}$ *are defined by*

$$v_j = \sum_i^{n_j} \beta_{j,i} B_{j,i}.$$

REMARK 2.2.6. *For a detailed discussion of the case where the conditions specified in Definitions* 2.2.3–2.2.5 *are satisfied in a finite element context, see Oswald* [70].

2.2.2. Besov norms.

Besov spaces were introduced in the late fifties and early sixties using Fourier decompositions as a means to classify functions and to measure their smoothness. In our context it is important that Besov spaces have an equivalent definition based on the moduli of smoothness, and that they be norm-equivalent to Sobolev spaces.

DEFINITION 2.2.6. *The* **Besov seminorm** $|\cdot|_{B^1}$ *is defined by*

$$|f|_{B^1} = \sqrt{\sum_{j=0}^{\infty} 2^{2j} \omega_2(2^{-j}, f)^2} \tag{2.35}$$

and the **Besov norm** $\|\cdot\|_{B^1}$ *is defined by*

$$\|f\|_{B^1} = \|f\|_{L_2} + |f|_{B^1}. \tag{2.36}$$

The **Besov space** $B^1(\Omega)$ *is defined as the set of functions*

$$B^1(\Omega) = \{f \in L_2(\Omega) |\; \|f\|_{B^1} < \infty\}. \tag{2.37}$$

Our definition of Besov norms is given in terms of finite differences, but our next theorem shows that they are equivalent to Sobolev norms. The finite difference interpretation of Besov norms makes them valuable for analyzing multilevel structures because their definition naturally links them to the multilevel setup. The Sobolev norms, on the other hand, are related to the differential operators and are less convenient for analyzing discrete techniques. The following basic result is taken from Triebel [104]; see also Oswald [70] and Dahmen and Kunoth [34].

THEOREM 2.2.1. *If Ω has the uniform cone property, then the Sobolev norm $\|\cdot\|_{H^1(\Omega)}$ is equivalent to the Besov norm $\|\cdot\|_{B^1(\Omega)}$: There exist constants $0 < c_1 \le c_2 < \infty$ such that, for all $u \in H^1(\Omega)$,*

$$c_1 \|u\|_{H^1(\Omega)} \le \|u\|_{B^1(\Omega)} \le c_2 \|u\|_{H^1(\Omega)}. \tag{2.38}$$

2.2.3. Upper bounds. We are now prepared to prove results for a multilevel splitting of the finite element space. The central result states that the approximation and inverse property imply a stable splitting. The proof consists of two parts, each formulated as a lemma.

We first exploit the *inverse property* to derive upper bounds for the H^1-norm in V.

LEMMA 2.2.1. *Assume that the inverse property* (2.32) *of Definition* 2.2.4 *holds in the form*

$$\omega_2(t, v_j) \le c(\min\{1, 2^{\lambda j} t^\lambda\}) \|v_j\|_{L_2}, \tag{2.39}$$

with $v_j \in V_j$ for $j \in \mathbb{N}_0$, where c is independent of v_j and j, and with $\lambda = 3/2$. Then there exists a constant $\bar{c}$ such that

$$\|u\|_{H^1}^2 \le \bar{c} \left(\|u_0\|_{H^1}^2 + \sum_{j=1}^{\infty} 2^{2j} \|u_j\|_{L_2}^2 \right) \tag{2.40}$$

for all $u \in H^1(\Omega)$ and for all decompositions of u with $u_j \in V_j$ and with

$$\sum_{j=0}^{\infty} u_j = u.$$

Proof. Lemma 2.2.1 is a special case of Theorem 4.1 of Dahmen and Kunoth [34], whose proof is based on a discrete Hardy inequality. In the following, we will give a direct proof, similar to Oswald [76, Thm. 2].

Because of the equivalence of the Sobolev space $H^1(\Omega)$ with the Besov space $B^1(\Omega)$, it suffices to show that

$$\|u\|_{B^1}^2 \le \bar{c} \left(\|u_0\|_{H^1}^2 + \sum_{j=1}^{\infty} 2^{2j} \|u_j\|_{L_2}^2 \right),$$

or simply that

$$|u|_{B^1}^2 = \sum_{k=0}^{\infty} 2^{2k} \omega_2(2^{-k}, u)^2 \le \bar{c} \sum_{j=0}^{\infty} 2^{2j} \|u_j\|_{L_2}^2. \tag{2.41}$$

From (2.39) we derive

$$\begin{aligned}\omega_2(2^{-k},u) &= \sum_{j=0}^{\infty}\omega_2(2^{-k},u_j)\\ &\le c\sum_{j=0}^{\infty}\min\{1,2^{\lambda(j-k)}\}\|u_j\|_{L_2}\\ &= c\left(\sum_{j=0}^{k}2^{\lambda(j-k)}\|u_j\|_{L_2}+\sum_{j=k+1}^{\infty}\|u_j\|_{L_2}\right).\end{aligned}$$

But then,

$$\begin{aligned}|u|_{B^1}^2 &= \sum_{k=0}^{\infty}2^{2k}\omega_2(2^{-k},u)^2\\ &\le c^2\sum_{k=0}^{\infty}2^{2k}\left(\sum_{j=0}^{k}2^{\lambda(j-k)}\|u_j\|_{L_2}+\sum_{j=k+1}^{\infty}\|u_j\|_{L_2}\right)^2. \qquad (2.42)\end{aligned}$$

Introducing a parameter $0<\epsilon$, by the Cauchy–Schwarz inequality the first term in the bound (2.42) satisfies

$$\begin{aligned}\left(\sum_{j=0}^{k}2^{\lambda(j-k)}\|u_j\|_{L_2}\right)^2 &= \left(\sum_{j=0}^{k}2^{\epsilon(j-k)}2^{-\epsilon(j-k)+\lambda(j-k)}\|u_j\|_{L_2}\right)^2\\ &\le \left(\sum_{j=0}^{k}2^{2\epsilon(j-k)}\right)\sum_{j=0}^{k}\left(2^{-\epsilon(j-k)+\lambda(j-k)}\|u_j\|_{L_2}\right)^2\\ &\le \frac{2^{2\epsilon}-2^{-2\epsilon k}}{2^{2\epsilon}-1}\sum_{j=0}^{k}\left(2^{-\epsilon(j-k)+\lambda(j-k)}\|u_j\|_{L_2}\right)^2.\end{aligned}$$

For all $k>0$, the factor in the last expression is bounded by a constant, depending only on ϵ, but not on k. Again the Cauchy–Schwarz inequality implies that the second term in (2.42) satisfies

$$\begin{aligned}\left(\sum_{j=k+1}^{\infty}\|u_j\|_{L_2}\right)^2 &= \left(\sum_{j=k+1}^{\infty}2^{-\epsilon j}2^{\epsilon j}\|u_j\|_{L_2}\right)^2\\ &\le \left(\sum_{j=k+1}^{\infty}2^{-2\epsilon j}\right)\sum_{j=k+1}^{\infty}2^{2\epsilon j}\|u_j\|_{L_2}^2\\ &\le \frac{4^{-\epsilon k-2\epsilon}}{2^{-2\epsilon}-2^{-4\epsilon}}\sum_{j=k+1}^{\infty}2^{2\epsilon j}\|u_j\|_{L_2}^2.\end{aligned}$$

With (2.42) and these two inequalities, and using

$$(a+b)^2\le 2(a^2+b^2)$$

for

$$a = \sum_{j=0}^{k} 2^{2j-2k} \|u_j\|_{L_2} \qquad \text{and} \qquad b = \sum_{j=k+1}^{\infty} \|u_j\|_{L_2},$$

we obtain

$$|u|_{B^1}^2 \le 2c^2 \sum_{k=0}^{\infty} 2^{2k} \left(\frac{2^{2\epsilon} - 2^{-2\epsilon k}}{2^{2\epsilon} - 1} \sum_{j=0}^{k} \left(2^{(\lambda-\epsilon)(j-k)} \|u_j\|_{L_2} \right)^2 + \frac{4^{-\epsilon k - 2\epsilon}}{2^{-2\epsilon} - 2^{-4\epsilon}} \sum_{j=k+1}^{\infty} 2^{2\epsilon j} \|u_j\|_{L_2}^2 \right).$$

Changing the order of summation, this becomes

$$\begin{aligned} |u|_{B^1}^2 &\le 2c^2 \sum_{j=0}^{\infty} \|u_j\|_{L_2}^2 \left(\sum_{k=0}^{j-1} 2^{2(k+\epsilon j)} \frac{4^{-\epsilon k - 2\epsilon}}{2^{-2\epsilon} - 2^{-4\epsilon}} + \sum_{k=j}^{\infty} 2^{2k} 2^{2(j-k)(\lambda-\epsilon)} \frac{2^{2\epsilon} - 2^{-2\epsilon k}}{2^{2\epsilon} - 1} \right) \\ &\le 2c^2 \sum_{j=0}^{\infty} \|u_j\|_{L_2}^2 \left(-\frac{16^{-\epsilon} \left(4^{\epsilon j} - 4^j\right)}{\left(2^{-2\epsilon} - 2^{-4\epsilon}\right)\left(4^{1-\epsilon} - 1\right)} + \frac{4^{j+2\epsilon} 4^{1-\lambda} - 4^{j+2\epsilon} + 4^{j-\epsilon j} 4^{\epsilon} - 4^{j-\epsilon j} 4^{1-\lambda+2\epsilon}}{\left(2^{2\epsilon} - 1\right)\left(4^{\epsilon} - 4^{1-\lambda+2\epsilon}\right)\left(4^{1-\lambda} - 1\right)} \right). \end{aligned}$$

Finally, for $\lambda = \frac{3}{2}$ and $\epsilon = \frac{1}{4}$, this reduces to

$$\begin{aligned} |u|_{B^1}^2 &\le 2c^2 \sum_{j=0}^{\infty} \|u_j\|_{L_2}^2 \frac{4^j - 2^{\frac{j}{2}} - 4 \cdot 2^{\frac{3j+1}{2}} + 3 \cdot 4^{j+\frac{1}{2}} + 4^{j+\frac{1}{2}}\sqrt{2} + 2^{\frac{3j+2}{2}}}{\left(\sqrt{2} - 1\right)\left(2\sqrt{2} - 1\right)} \\ &= \frac{2c^2}{(\sqrt{2} - 1)} \sum_{j=0}^{\infty} \|u_j\|_{L_2}^2 \left(\frac{7 + 2\sqrt{2}}{\left(2\sqrt{2} - 1\right)} 2^{2j} - 2 \cdot 2^{\frac{3}{2} j} - \frac{1}{\left(2\sqrt{2} - 1\right)} 2^{\frac{1}{2} j} \right). \end{aligned}$$

This shows (2.41) and thus proves (2.40) and the lemma.

REMARK 2.2.7. *Choosing $\epsilon \ne \frac{1}{4}$ in the proof of Lemma 2.2.1 may allow for sharper bounds.*

The upper bound of Lemma 2.2.1 holds for an arbitrary decomposition of $u \in V$. Taking the infimum of all splittings we derive the following corollary.

COROLLARY 2.2.1. *Under the hypothesis of Lemma 2.2.1, there exists a constant $\bar{c}$ such that*

$$\|u\|_{H^1} \le \bar{c} |||u|||. \tag{2.43}$$

Proof. Lemma 2.2.1 implies

$$(2.44)\ \|u\|_{H^1} \le \bar{c}\ \inf\left\{ \left(\|u_0\|_{H^1}^2 + \sum_{j=1}^{\infty} 2^{2j}\|u_j\|_{L_2}^2 \right)^{\frac{1}{2}} \;\middle|\; \begin{array}{c} u_j \in V_j \text{ for } j \in \mathbb{N}_0, \\ \sum_{j=0}^{\infty} u_j = u \end{array} \right\}.$$

The right-hand side is just the definition of the additive Schwarz norm $|||\cdot|||$ (cf. (2.4)).

REMARK 2.2.8. *The upper bound* (2.40) *or* (2.43) *is an implication of the inverse property* (2.32).

REMARK 2.2.9. *Clearly,* (2.40) *and* (2.43) *can be used to derive error estimates. We defer a discussion until* §2.6.

REMARK 2.2.10. *In Bornemann and Yserentant* [20], *the upper bound is derived for linear finite elements without using the theory of Besov spaces. The core of their derivation is the proof of a so-called* strengthened Cauchy–Schwarz inequality.

2.2.4. Lower bounds. Having derived bounds for $\|\cdot\|_{H^1}$ from above, we now derive lower bounds.

LEMMA 2.2.2. *If the approximation property* (*see Definition* 2.2.3) *holds for the collection of subspaces* $\{V_j\}_{j\in\mathbb{N}_0}$ *of* $V = H^1(\Omega)$ *in the form*

$$E_j(u) \le \hat{c}\omega_2(2^{-j}, u) \tag{2.45}$$

(*see also Remark* 2.2.3), *then there exists a constant* $0 < \tilde{c} < \infty$, *such that*

$$\|u\|_{H^1} \ge \tilde{c}|||u||| \tag{2.46}$$

for all $u \in H^1(\Omega)$.

Proof. Let $u \in H^1(\Omega)$ be given and consider its L_2*-best approximations* $u_j^* \in V_j$, defined by

$$\|u - u_j^*\|_{L_2} = E_j(u)$$

for all $j \in \mathbb{N}_0$. With $\tilde{u}_j$ defined by

$$\tilde{u}_j = u_j^* - u_{j-1}^*$$

for $j \in \mathbb{N}$, and $\tilde{u}_0 = u_0^*$, we have a decomposition of u given by

$$u = \sum_{j=0}^{\infty} \tilde{u}_j. \tag{2.47}$$

For this decomposition, by (2.45) we have

$$\sum_{j=1}^{\infty} 2^{2j}\|\tilde{u}_j\|_{L_2}^2 \le 2\sum_{j=1}^{\infty} 2^{2j}(\|u - u_j^*\|_{L_2}^2 + \|u - u_{j-1}^*\|_{L_2}^2)$$

$$\begin{aligned}
&\leq 10 \sum_{j=0}^{\infty} 2^{2j} \|u - u_j^*\|_{L_2}^2 \\
&\leq 10 \sum_{j=0}^{\infty} 2^{2j} E_j(u)^2 \\
&\leq 10\hat{c}^2 \sum_{j=0}^{\infty} 2^{2j} \omega_2(2^{-j}, u)^2 \\
&\leq 10\hat{c}^2 |u|_{B^1}^2 \\
&\leq c_1 \|u\|_{H^1}^2,
\end{aligned}$$

where $0 < c_1 < \infty$ is a suitable constant. Furthermore, there exists a constant $\bar{c}$ such that

$$\begin{aligned}
\|\tilde{u}_0\|_{H^1}^2 &= \|u + \tilde{u}_0 - u\|_{H^1}^2 \\
&\leq 2\|u\|_{H^1}^2 + 2\|u - \tilde{u}_0\|_{H^1}^2 \\
&\leq \bar{c} \left(\|u\|_{H^1}^2 + \sum_{j=1}^{\infty} 2^{2j} \|\tilde{u}_j\|_{L_2}^2 \right).
\end{aligned}$$

The last inequality was obtained by applying Lemma 2.2.1 to

$$u - \tilde{u}_0 = \sum_{j=1}^{\infty} \tilde{u}_j.$$

Thus, the $\tilde{u}_j$, $j \in \mathbb{N}_0$, provide a decomposition of u such that, for some constant $0 < \tilde{c} < \infty$,

$$\|\tilde{u}_0\|_{H^1}^2 + \sum_{j=1}^{\infty} 2^{2j} \|\tilde{u}_j\|_{L_2}^2 \leq \frac{1}{\tilde{c}} \|u\|_{H^1}^2.$$

To get the additive Schwarz norm, we consider the infimum of all decompositions to obtain (2.46), which proves the lemma.

REMARK 2.2.11. *Lower bounds of the form*

$$\sum_{j=0}^{\infty} 2^{2j} \|\tilde{u}_j\|_{L_2}^2 \leq \frac{1}{c} \|u\|_{H^1},$$

where $\tilde{u}$ is estimated in L_2, can also be obtained without using Lemma 2.2.1. The lower bound is an implication of the approximation property (2.30) alone.

2.2.5. Main theorem. We are now ready to prove the central result of this chapter.

THEOREM 2.2.2. *Assume that we have a nested system of C^0-finite element spaces $\{V_j\}_{j \in \mathbb{N}_0}$ generated by a sequence of uniform refinements of an initial regular triangulation of a plane, polygonal domain $\Omega \subset \mathbb{R}^2$ satisfying the* *uniform cone condition such that*

$$\sum_{j=0}^{\infty} V_j = V \equiv H^1(\Omega).$$

Let the Hilbert space structure of the spaces V_j be defined by

$$(2.48) \qquad (u,v)_{V_j} = \begin{cases} \langle u,v\rangle_V, & \text{for } j=0, \\ 2^{2j}(u,v)_{L_2(\Omega)}, & \text{for } j \geq 1, \end{cases}$$

where $\langle\cdot,\cdot\rangle_V$ denotes the H^1-inner product. Then $\{V_j\}_{j\in\mathbb{N}_0}$ forms a stable splitting of $V = H^1(\Omega)$.

Proof. In Oswald [70] it is shown that the system of spaces $\{V_j\}_{j\in\mathbb{N}_0}$ satisfies the approximation and inverse property, as required to apply Lemma 2.2.2 and Corollary 2.2.1. The two together directly yield the bounds required for a stable splitting.

2.3. Stable bases

In this section the stable splitting in a multilevel hierarchy will be extended using the concept of a *stable basis*, as introduced in Definition 2.2.5. Using the bases $\mathcal{B}_j = \{B_{j,i} | i = 1, 2, \ldots, n_j\}$, we define the one-dimensional spaces

$$V_{j,i} = \text{span}(B_{j,i}), \qquad i = 1,2,3,\ldots,n_j \quad \text{for } j \in \mathbb{N}.$$

According to Definition 2.2.5, we introduce in $V_{j,i}$ the inner product

$$(u,v)_{V_{j,i}} = d_j 2^{2j}\alpha\beta$$

for $u = \alpha B_{j,i} \in V_{j,i}$ and $v = \beta B_{j,i} \in V_{j,i}$, $j \in \mathbb{N}$, and $i = 1, 2, \ldots, n_j$. The corresponding norm is defined by $\|\cdot\|_{V_{j,i}} = \sqrt{(\cdot,\cdot)_{V_{j,i}}}$. The special role of the coarsest space is taken into account by defining $n_0 = 1$ and $V_{0,1} = V_0$, and by using the norm $\|\cdot\|_{V_{0,1}} \equiv \|\cdot\|_{V_0}$.

THEOREM 2.3.1. *If the collection of bases $\{\mathcal{B}_j\}_{j\in\mathbb{N}_0}$ of $\{V_j\}_{j\in\mathbb{N}_0}$ is stable (see Definition 2.2.5), then the collection of one-dimensional spaces*

$$\{V_{j,i}\}_{j\in\mathbb{N}_0;\, i=1,2,\ldots,n_j}$$

forms a stable splitting of $V = H^1(\Omega)$:

$$H^1(\Omega) = V_0 + \sum_{j=1}^{\infty}\sum_{i=1}^{n_j} V_{j,i}.$$

Proof. Each $u_j \in V_j$ uniquely decomposes with respect to the basis $\mathcal{B}_j$,

$$u_j = \sum_{i=1}^{n_j} u_{j,i} B_{j,i},$$

with real coefficients $u_{j,i}$ for $j \in \mathbb{N}$, $i = 1, \ldots, n_j$. From Definition 2.2.5 we know that there exist constants d_j and $\bar{c}$ such that

$$\|u_j\|_{L_2}^2 \leq \bar{c} d_j \sum_{i=1}^{n_j} u_{j,i}^2.$$

Lemma 2.2.1 shows that for any decomposition of $u \in V$ by $u_j \in V_j$ for $j \in \mathbb{N}_0$ with $u = \sum_{j=0}^{\infty} u_j$, we get the estimate

$$\begin{aligned} \|u\|_{H^1}^2 &\leq \hat{c} \left(\|u_0\|_{H^1}^2 + \sum_{j=1}^{\infty} 2^{2j} \|u_j\|_{L_2}^2 \right) \\ &\leq \hat{c} \left(\|u_0\|_{H^1}^2 + \sum_{j=1}^{\infty} 2^{2j} \bar{c} d_j \sum_{i=1}^{n_j} u_{j,i}^2 \right) \\ &\leq \tilde{c} \left(\|u_0\|_{H^1}^2 + \sum_{j=1}^{\infty} \sum_{i=1}^{n_j} \|u_{j,i} B_{j,i}\|_{V_{j,i}}^2 \right). \end{aligned}$$

This holds for any decomposition of u, so by taking the infimum we get (the triple bar norm $|||\cdot|||$ is now to be understood with respect to the refined splitting)

$$\|u\|_{H^1}^2 \leq \tilde{c} |||u|||^2 \equiv \inf \left\{ \|u_0\|_{H^1}^2 + \sum_{j=1}^{\infty} \sum_{i=1}^{n_j} \|u_{j,i} B_{j,i}\|_{V_{j,i}}^2 \right| \left. u = u_0 + \sum_{j=1}^{\infty} \sum_{i=1}^{n_j} u_{j,i} B_{j,i};\, u_0 \in V_0 \right\}.$$

The infimum can be extended to cover all decompositions in the refined splitting because the decomposition of $u_j = \sum_{i=1}^{n_j} u_{j,i} B_{j,i}$ is unique. This establishes the required upper bound.

To prove the the lower bound we observe that

$$\begin{aligned} \|u\|_{H^1}^2 &\geq \bar{c} \inf \left\{ \|u_0\|_{H^1}^2 + \sum_{j=1}^{\infty} 2^{2j} \|u_j\|_{L_2}^2 \,\middle|\, \sum_{j=0}^{\infty} u_j = u;\, u_j \in V_j \right\} \\ &\geq \bar{c} \inf \left\{ \|u_0\|_{H^1}^2 + \sum_{j=1}^{\infty} 2^{2j} \hat{c} d_j \sum_{i=1}^{n_j} u_{j,i}^2 \,\middle|\, \sum_{j=0}^{\infty} u_j = u;\, u_j \in V_j \right\} \\ &\geq \bar{c} \inf \left\{ \|u_0\|_{H^1}^2 + \sum_{j=1}^{\infty} 2^{2j} \hat{c} d_j \sum_{i=1}^{n_j} u_{j,i}^2 \,\middle|\, u_0 + \sum_{j=0}^{\infty} \sum_{i=1}^{n_j} u_{j,i} B_{j,i} = u \right\} \\ &\geq \tilde{c} |||u|||^2. \end{aligned}$$

This concludes the proof of the second inequality and thus completes the proof of the lemma.

2.4. Induced splittings

For practical applications we must consider finite-dimensional spaces.

THEOREM 2.4.1. *A nested stable splitting* $\{V_j\}_{j\in\mathbb{N}_0}$ *with* $V_0 \subset V_1 \subset \cdots \subset V$ *of* $V = H^1(\Omega)$ *induces a stable splitting of any of its subspaces* $(V_K, \|\cdot\|_{H^1})$ (V_K, *equipped with the* H^1-*norm*). *For these splittings the stability constants are bounded uniformly in* $K \in \mathbb{N}$.

Proof. Clearly,

$$V_K = \sum_{j=0}^{K} V_j. \tag{2.49}$$

For this proof we must show that the proofs of Lemmas 2.2.1 and 2.2.2 remain valid when the splitting is finite and, correspondingly, that the infimum is taken over the finite collection of subspaces.

The upper bound is straightforward because

$$\begin{aligned}\|u_K\|_{H^1} &\le c_0 \inf\left\{ \|u_0\|_{H^1} + \sum_{j=0}^{\infty} 2^{2j}\|u_j\|_{L_2} \,\middle|\, \sum_{j=0}^{\infty} u_j = u \right\} \\ &\le c_0 \inf\left\{ \|u_0\|_{H^1} + \sum_{j=0}^{K} 2^{2j}\|u_j\|_{L_2} \,\middle|\, \sum_{j=0}^{K} u_j = u \right\}.\end{aligned}$$

To prove the lower bound we use the decomposition of u_K by differences of best approximations as in (2.47). However, because $u_j^* = u_K$ for all $j \ge K$, we have that $\tilde{u}_j = 0$ whenever $j > K$. Thus, we have a suitable decomposition of u_K with

$$\sum_{j=0}^{K} 2^{2j}\|\tilde{u}_j\|_{L_2} \le c\|u_K\|_{H^1},$$

and we may on the left side take the infimum of all decompositions with

$$u_K = \sum_{j=0}^{K} \hat{u}_j,$$

where $\hat{u}_j \in V_j$ for $j = 0, 1, \ldots, K$.

This result clearly extends to the case of a refined splitting with respect to a stable basis.

COROLLARY 2.4.1. *The (finite) collection of spaces*

$$\{V_{j,i}\}_{j=1,\ldots,K;i=1,\ldots,n_j}$$

forms a stable splitting of $(V_K, \|\cdot\|_{H^1})$ *with stability constants bounded uniformly in* K.

2.5. Multilevel iterations

The theory developed in §§2.1–2.4 may now be used to give a foundation for iterative algorithms. In §2.1, the concept of a stable splitting has led us to reformulate the abstract problem (2.2) in an equivalent form (2.18). The operator in this equivalent form is shown to have bounded spectrum.

The convergence behavior for some iterative methods is directly determined by the condition number, that is, the relation between largest and smallest eigenvalues. Such methods, most notably *steepest descent* and *conjugate gradient iteration* (see Axelsson and Barker [2]), can be applied directly or indirectly to the operator equation (2.18) and estimates of the condition number give bounds on the rate of convergence.

For an efficient implementation of iterative methods, an explicit construction of P_V must be avoided. Fortunately, many iterative methods do not need the operator explicitly, but require only the repeated application of the operator to a vector. The details of an efficient implementation will be discussed in Chapter 3, with special emphasis on the combination of our techniques with adaptive mesh refinement.

For other iterative schemes, like *Gauß–Seidel iteration* or *successive overrelaxation* (SOR) and its variants, the convergence theory is not based on condition numbers alone. Estimates for the speed of convergence are less direct. In Chapter 3, we will propose and analyze a generalization of the Gauß–Southwell method that is especially designed for the adaptive refinement case.

2.6. Multilevel error estimators

Besides providing the theoretical basis for the fast iterative solution of discretized PDEs, the multilevel splittings can also be used to provide error estimates (see Remark 2.2.9). The inequalities (2.40) and (2.43) can clearly be reinterpreted as bounds for the error. The results are stated in the two theorems below.

THEOREM 2.6.1. *Assume that the collection of spaces $\{V_j\}_{j\in J}$ is a stable splitting of V with*

$$c_1 \le \lambda_{\min}(P_V) \le \lambda_{\max}(P_V) \le c_2, \tag{2.50}$$

and that u^* is the solution of (2.2) *in V. Then, for any Hilbert space norm $\|\cdot\|$ in V and all $u \in V$, we have*

$$\frac{1}{c_2}\left\|\sum_{j\in J}(P_{V_j}u-\phi_j)\right\| \le \|u-u^*\| \le \frac{1}{c_1}\left\|\sum_{j\in J}(P_{V_j}u-\phi_j)\right\|. \tag{2.51}$$

Proof. This is a direct consequence of the bounds on the spectrum (2.50) and the relation

$$P_{V_j}u^* = \phi_j,$$

which follows from Definition 2.1.3.

REMARK 2.6.1. *The value of Theorem* 2.6.1 *is that it estimates an unknown quantity, the error* $u - u^*$, *by known quantities, the* residuals $P_{V_j}u - \phi_j$. *For an algebraic interpretation and applications, see Chapter* 3.

For our next theorem, we introduce the *scaled residuals.*

DEFINITION 2.6.1. *The* **scaled residuals** $\bar{r}_j \in V_j$ *of* u *are defined by*

$$b_j(\bar{r}_j, v_j) = a(u - u^*, v_j) = a(u, v_j) - \Phi(v_j) \tag{2.52}$$

for all $v_j \in V_j$, $j \in J$, *where* u^* *is the solution of* (2.2) *in* V.

THEOREM 2.6.2. *Assume that the collection of spaces* $\{V_j\}_{j\in J}$ *is a stable splitting of* V *and that* u^* *is the solution of* (2.2) *in* V. *Then there exist constants* $0 < c_0 \leq c_1 < \infty$ *such that*

$$c_0 \sum_{j\in J} \|\bar{r}_j\|^2_{V_j} \leq \|u - u^*\|^2_V \leq c_1 \sum_{j\in J} \|\bar{r}_j\|^2_{V_j}. \tag{2.53}$$

Proof. With inequalities (2.1) and equation (2.13) it suffices to show that there exist constants $0 < \bar{c}_0 \leq \bar{c}_1 < \infty$ such that

$$\bar{c}_0 \sum_{j\in J} b_j(\bar{r}_j, \bar{r}_j) \leq a(u - u^*, u - u^*) \leq \bar{c}_1 \sum_{j\in J} b_j(\bar{r}_j, \bar{r}_j). \tag{2.54}$$

From Theorem 2.1.1, we know that there exist such constants with

$$\bar{c}_0 a(P_V(u - u^*), u - u^*) \leq a(u - u^*, u - u^*) \leq \bar{c}_1 a(P_V(u - u^*), u - u^*).$$

Additionally,

$$\begin{aligned} a(P_V(u - u^*), u - u^*) &= \sum_{j\in J} a(u - u^*, P_{V_j}(u - u^*)) \\ &= \sum_{j\in J} b_j(\bar{r}_j, P_{V_j}(u - u^*)) \\ &= \sum_{j\in J} a(\bar{r}_j, (u - u^*)) \\ &= \sum_{j\in J} b_j(\bar{r}_j, \bar{r}_j), \end{aligned}$$

which concludes the proof.

REMARK 2.6.2. *The error estimate in Theorem* 2.6.2 *provides bounds in the* natural (energy) norms. *This is different from Theorem* 2.6.1, *where any norm can be used, and where the bounds are somewhat different.*

The error estimates in this section are a natural consequence of the multilevel structure. In contrast to conventional error estimators, they use residuals on not just one, but on all levels. The same structure that provides fast iterative solvers can thus be used to calculate sharp, two-sided error bounds. The practical application of these error estimates is discussed in Chapter 3.

Conventional error estimators only employ a single level, say $j^* \in J$. This can be linked to our multilevel estimates by demanding that the residuals $P_{V_j}u - \phi_j = 0$ vanish for all other levels $j \in J, j \neq j^*$. Then the contributions of these levels vanish in the multilevel sums (2.51) and (2.54), and we get an estimate based on one level only.

In practice, $J = \{0, 1, 2, \dots, j^*\}$ and j^* will be the finest level. If the residuals on all levels vanish, except on level j^*, this is equivalent to solving the coarser levels exactly. On the finest level, we have the interpolant of the solution from the next coarser level. Here the residuals on the finest level alone provide an estimate for the error. Note that for a finite-dimensional problem (J finite), this estimate bounds the *algebraic error*, that is, the difference between the fine and the coarse level solution.

Of course, being able to estimate the difference of a fine to a coarse mesh solution is already a useful *error indicator* to guide the refinement process heuristically. We wish to point out that for the case of linear triangular elements, this simple error indicator can be shown to be equivalent to error estimators based on introducing higher order basis functions, as the *edge-oriented* error estimator of KASKADE (see Deuflhard, Leinen, and Yserentant [35]). This is shown by using extrapolation arguments in Rüde [85], [90].

However, the error bounds (2.51) and (2.54) also apply in the case J infinite, say for $J = \mathbb{N}_0$, corresponding to the case of successive refinement of a finite element triangulation. If we again assume $P_{V_j}u - \phi_j = 0$ for $j < j^*$, then the error bounds are obtained by a sum of residuals of the form $\sum_{\nu=j^*}^{\infty}$, that is, all levels finer than the current one. Note that this expression now provides, at least formally, an estimate for the *discretization error* (differential error). Clearly, the infinite sum cannot be evaluated directly. However, for *sufficiently regular* solutions, the residuals on level j^* can be used to bound the residuals on finer levels $j > j^*$.

For piecewise quadratic (on level j^*) solutions, we may assume that the residuals decrease with a factor $1/4^j$ for $j > j^*$, so that the contributions of the levels to the error bound become a geometric sum that can be evaluated. This leads to an error estimator that is again closely related to the error estimators used in KASKADE or PLTMG (see [7]). So called **hierarchical basis error estimators** are also discussed in Verführt [106]. Equivalent relations for the residuals on finer levels can be found, if we require bounds on higher derivatives of the solution. With this type of assumption we will only need to evaluate the residuals on a single level j^* to estimate the residual of all finer levels and thus obtain a bound for the discretization error.

Chapter 3

The Fully Adaptive Multigrid Method

In contrast to most other iterative elliptic solvers the *fully adaptive multigrid method* (FAMe) uses strategies to exploit both the matrix structure and the current state of the iteration. The core of the method consists of a relaxation scheme supplemented with an *active set strategy.* The active sets, which are used to monitor where the iteration efficiently reduces the error, are updated *incrementally.* This is done by the current approximate solution and the matrix structure. Arithmetic operations are restricted to the active set, so that computational work will only be spent for unknowns for which the error norm can be efficiently reduced.

The effectiveness of the method as a solver depends on whether it can be combined with preconditioning techniques based on a multilevel structure. The active set strategy can be extended to this structure by additionally tracing the dependencies between unknowns on different levels. The theory of multilevel algorithms, particularly the results developed in Chapter 2, can be adapted to show that the resulting *multilevel adaptive iteration* has an asymptotically optimal convergence rate where this theory applies.

The main advantage of the method is that it improves on the robustness and efficiency of classical multilevel methods for nontrivial problems; in particular, it is an almost ideal supplement to (self-)adaptive refinement techniques. With suitable data structures, the overhead can be kept small, at most proportional to the other numerical work.

Classical multilevel adaptive methods are adaptive in the mesh generation, but the solution algorithm is fixed. Consider the case of a local refinement being added to a global grid after the coarse grid problem has been solved. The equations in the refined region must now be solved. It is important to realize that it is not sufficient to solve the equations on the new level alone. It is necessary to return to the coarse meshes to compute any global effects that may originate from the locally improved resolution. Various types of multilevel algorithms accomplish this in the correct way.

Assume that an initial guess for the values at the new nodes has been computed by interpolation. The iteration to improve this solution first consists of relaxation and coarsening only, both of which are local operations (except for convection dominated equations). Therefore, all significant changes are restricted to the newly refined region and any effects of the refinement are restricted to a neighborhood of the refined region. Only in the interpolation of corrections from coarse to fine meshes do global effects begin to be felt.

Conventional multigrid implementations, however, would relax all unknowns on the coarser levels, independently of whether they can be effected by the refinement or not. Consequently, a classical implementation of multigrid will waste many operations, because it does not exploit the deferred spreading of information through the region. Adaptive codes attempt to limit this effect by prescribing a certain minimal size for the refinement (in number of new nodes relative to the existing mesh). The FAMe technique is designed to exploit this character of the refinement process more systematically and thereby is an improvement over conventional multilevel methods.

The robustness of multigrid methods can be generally improved by so-called local relaxations; see Brandt [30], Mikulinsky [63], and Stevenson [101]. It has commonly been observed that the multigrid rate of convergence (in particular, for V-cycles) depends sensitively on effects like singularities or perturbations from boundary conditions. This sensitivity can be ameliorated by adding extra relaxation sweeps that are confined to a neighborhood (a small number of surrounding grid points) of the perturbation. The multilevel adaptive relaxation technique will automatically take care of these situations as well.

The presentation in this chapter will be algebraic in order to emphasize that the method can be applied to a wide class of sparse linear systems. The presently available theory, however, depends on the abstract theory of Chapter 2, and thus the properties of finite elements for second order elliptic PDEs. For an example where multilevel algorithms are successfully applied to a problem in Chip layout design that does not arise from a discretization process, see Regler and Rüde [79].

We will start by presenting the *adaptive relaxation technique* as an extension of either a Gauß–Seidel-like or a Jacobi-like iteration. Both alternatives can be analyzed theoretically. For the Jacobi-like relaxation we use perturbation arguments relating the algorithm to classical multigrid theory. The Gauß–Seidel-like relaxation scheme can be analyzed directly with results in Chapter 2 by examining its elementary steps and showing that for each of them the error is reduced substantially.

Adaptive iteration fits especially well into the context of adaptive mesh refinement with the virtual global grid technique, introduced in §3.7 (see also Chapter 4). The virtual global grids represent the sequence of uniformly refined grids as an (*infinite*) *recursive data structure*. The construction of the structure, however, is implemented by *lazy evaluation*, so that in practice only a finite subset of the structure needs to be built.

The error bound that provides stopping criteria for the iteration can also serve as an error indicator for this specific mesh refinement technique. Discrete and differential errors are interpreted and analyzed with the same theoretical device. This gives a dual view of the FAMe method: The FAMe can be seen as a solver plus mesh refinement and mesh generation, and as a solver for an infinite sequence of algebraic equations that are all solved to within a specified tolerance.

Before we discuss these concepts for the multilevel case, we introduce the *active set concept* and *adaptive relaxation technique* for linear systems.

3.1. Adaptive relaxation

In this section, we will discuss two variants of *adaptive relaxation*, which is an extension of classical relaxation techniques that are the basis for the FAMe. We discuss the algorithm in its interpretation as an iterative linear system solver. In the subsequent sections, the adaptive relaxation technique will be generalized and will become the core component of multilevel adaptive iteration schemes.

The discretization of equation (1.1) or (1.2) yields a sparse, symmetric, linear system of equations,

$$Ax = b, \tag{3.1}$$

where

$$\begin{aligned} A &= (a_{i,j})_{1\leq i,j\leq n} \in \mathbb{R}^{n\times n}, \\ x &= (\xi_i)_{1\leq i\leq n} \in \mathbb{R}^n, \\ b &= (b_i)_{1\leq i\leq n} \in \mathbb{R}^n. \end{aligned} \tag{3.2}$$

We assume that A is symmetric, positive definite, such that the solution of (3.1) is equivalent to solving the minimization problem

$$\min_{x\in\mathbb{R}^n} (x^T Ax - 2x^T b). \tag{3.3}$$

The adaptive relaxation algorithm makes explicit use of the graph associated with the matrix. In particular, we will need the following definition.

DEFINITION 3.1.1. *For each index* $1 \leq i \leq n$, *we define the set of* **connections** *by*

$$\text{Conn}(i) = \{j|\ j \neq i, a_{j,i} \neq 0\}. \tag{3.4}$$

REMARK 3.1.1. Conn(i) *is the set of indices corresponding to the nonzero entries in* column i *of the system matrix, excluding the diagonal (see also definition* (4.10)). *Because of the symmetry of* A, Conn(i) *is also the set of indices corresponding to the nonzero entries in* row i, *excluding the diagonal.*

REMARK 3.1.2. *We assume that* A *is* sparse, *that is, that the number of connections is small:*

$$|\text{Conn}(i)| \leq \bar{C} \ll n \tag{3.5}$$

for all i, $1 \leq i \leq n$.

3.1.1. Sequential adaptive relaxation.

DEFINITION 3.1.2.

$$\theta_i = \theta_i(x) = a_{i,i}^{-1} e_i^T (b - Ax)$$

is called the **current scaled residual** *of equation i, where e_i is the ith unit vector, that is, the ith column of the unit matrix.*

REMARK 3.1.3. *$\theta_i(x)$ is also called the* dynamic residual, *which plays an important role in analyzing multigrid methods; see Brandt* [29].

Based on the scaled residuals we may introduce *relaxation* in the following form.

DEFINITION 3.1.3. *An* **elementary relaxation step** *for equation i consists of a change of vector x by*

$$x' \longleftarrow x + \theta_i(x) e_i. \tag{3.6}$$

REMARK 3.1.4. *An elementary relaxation step for equation i can be understood as a coordinate descent for the ith coordinate direction, and the scaled residual is the solution of the one-dimensional minimization problem*

$$\min_{\zeta \in \mathbb{R}} ((x + \zeta e_i)^T A (x + \zeta e_i) - 2(x + \zeta e_i)^T b), \tag{3.7}$$

originating from projecting the functional (3.3) *to the ith coordinate direction.*

REMARK 3.1.5. *Written explicitly, an elementary relaxation step becomes*

$$\xi_i \longleftarrow a_{i,i}^{-1} \left(b_i - \sum_{j \in \text{Conn}(i)} a_{i,j} \xi_j \right). \tag{3.8}$$

Many iterative solvers are based on different combinations of elementary relaxation steps. The Gauß–Seidel method simply performs n basic relaxation steps in some predetermined order; the Gauß–Southwell algorithm (see Luenberger [53] and Southwell [96], [98], [99]) performs elementary relaxation steps for the equation whose current residual has the largest absolute value.

To evaluate the quality of an approximate solution, we will use either the Euclidean norm, the maximum norm, or the *energy* norm defined by

$$\begin{aligned} \|x\|_2 &= \sqrt{\frac{1}{n} \sum_{i=1}^{n} \xi_i^2}, \\ \|x\|_\infty &= \max_{i=1,\dots,n} |\xi_i|, \\ \|x\|_E &= \sqrt{x^T A x}, \end{aligned} \tag{3.9}$$

respectively. The norms are used to measure the error $x - x^*$ of an approximate solution x, where

$$x^* = A^{-1} b$$

is the exact solution. We also introduce the algebraic analogue of the finite element residual in Definition 2.6.1.

DEFINITION 3.1.4. *The* **residual** *of a vector x is defined as*

$$r = b - Ax, \tag{3.10}$$

and the **scaled residual** *is defined as*

$$\bar{r} = D^{-1}(b - Ax), \tag{3.11}$$

where $D = \operatorname{diag}(A)$ is the diagonal part of A.

REMARK 3.1.6. *The* scaled residual $\bar{r}$ *and the* scaled residual of equation i *are related by*

$$\theta_i = e_i^T \bar{r} \tag{3.12}$$

and

$$\bar{r} = \sum_{i=1}^{n} \theta_i e_i. \tag{3.13}$$

Elementary relaxation steps reduce the error of an approximate solution. This effect can be described by the following lemma.

LEMMA 3.1.1. *Let $x - x^*$ be the error before, and $x' - x^*$ be the error after, an elementary relaxation step of the form* (3.6) *for equation i. Then the energy norm of the error is reduced according to*

$$\|x - x^*\|_E^2 - \|x' - x^*\|_E^2 = a_{i,i}\theta_i(x)^2. \tag{3.14}$$

Proof.

$$\begin{aligned} &\|x - x^*\|_E^2 - \|x' - x^*\|_E^2 \\ &= (x - x^*)^T A(x - x^*) - (x + \theta_i e_i - x^*)^T A(x + \theta_i e_i - x^*) \\ &= -2\theta_i e_i^T A(x - x^*) - \theta_i^2 e_i^T A e_i \\ &= -2\theta_i(-a_{i,i}\theta_i) - a_{i,i}\theta_i^2 \\ &= a_{i,i}\theta_i^2. \end{aligned}$$

Note that A being positive definite implies $a_{i,i} > 0$.

Lemma 3.1.1 shows that the error reduction is fast when and only when equations with large residuals (relative to $\sqrt{a_{i,i}}$) are relaxed. This is exploited in the method of Gauß–Southwell, where repeatedly the equation with *largest* component in the gradient (i.e., where $a_{i,i}\theta_i$ is largest) is selected for an elementary relaxation step. The Gauß–Southwell method is often too expensive in practical implementations, because it requires the determination of the maximal residual in each step. In general, this requires the evaluation of n residuals, if no incremental update strategy is used.

In our algorithm, we therefore use weaker conditions. We classify the residuals according to their magnitude and introduce sets of equations with *large* residuals.

DEFINITION 3.1.5. *The* **strictly active set** $S(\theta, x)$ *is defined by*

$$S(\theta, x) \stackrel{\text{def}}{=} \{i| \; |\theta_i(x)| \geq \theta\}. \tag{3.15}$$

REMARK 3.1.7. *In view of Lemma* 3.1.1 *the active set could alternatively be defined by*

$$S(\theta, x) \stackrel{\text{def}}{=} \{i| \; |a_{i,i}\theta_i(x)| \geq \theta\}$$

or

$$S(\theta, x) \stackrel{\text{def}}{=} \{i| \; |\sqrt{a_{i,i}}\theta_i(x)| \geq \theta\}.$$

This is uncritical if we assume that all $a_{i,i}$ *have approximately the same size. In a multilevel setup, however, the* $a_{i,i}$ *may be associated with different levels and their size may depend on the mesh level. We will use residuals of all levels of the mesh structure for error estimates and will see that the scaling of the residuals leads to different error estimators; see Corollaries* 3.3.2 *and* 3.3.3 *below.*

REMARK 3.1.8. *If relaxation could be restricted to the strictly active set for some tolerance* θ, *Lemma* 3.1.1 *would guarantee a lower bound on the reduction of the energy norm of the error in each elementary relaxation step.*

Unfortunately, the sets $S(\theta, x)$ are still too expensive to compute exactly. Therefore, instead of working with $S(\theta, x)$ directly, the *sequential adaptive relaxation* algorithm in Fig. 3.1 uses *extended* active sets $\tilde{S}(\theta, x)$.

DEFINITION 3.1.6. $\tilde{S}(\theta, x)$ *is called an* **active set** *if*

$$S(\theta, x) \subset \tilde{S}(\theta, x) \subset \{1, 2, \ldots, n\}. \tag{3.16}$$

Our scheme for determining $\tilde{S}$ must balance two conflicting goals. Clearly, error reduction will be more efficient if $\tilde{S}$ is closer to S. On the other hand, an exact determination of S is too expensive and must be avoided. For a single mesh level, a scheme that is properly balanced is given in Fig. 3.1. The main loop contains an elementary relaxation step and an update of the active set. An active set will generally also contain indices of equations with small residuals. This must be taken into account when relaxing the members of the extended active set.

The key idea of the adaptive relaxation strategy is

Do not update unknowns with small scaled residual.

This gives rise to an efficient scheme for updating the active set. For a graphical illustration of a basic step of the sequential adaptive iteration, see Fig. 3.2.

At a particular node, an improved value is computed by calculating the local solution of the equation. For this step, the values of neighboring nodes must be collected. The node is updated if the new value differs from the old value by more than a minimal amount. All neighboring nodes are then added to the active set. If the new value does not differ much from the old one, the update is skipped, the node keeps its old value, and the neighbors keep

```
proc SequentialAdaptiveRelaxation( θ, x, S̃ )
    assert( θ > 0 )
    assert( S̃ ⊃ S(θ, x) )
    while( S̃ ≠ ∅ )
        pick i ∈ S̃
        S̃ ← S̃ \ {i}
        if |θ_i(x)| > θ then
            x ← x + θ_i(x)e_i
            S̃ ← S̃ ∪ Conn(i)
        end if
        assert( S̃ ⊃ S(θ, x) )
    end while
    assert( S̃ = ∅ )
end proc
```

FIG. 3.1. *Sequential adaptive relaxation.*

their active/inactive state. In any case, the current node can be deleted from the active set because another relaxation would yield exactly the same result. The node can, of course, again become active if its neighbors change in later relaxation steps.

The input for the algorithm in Fig. 3.1 is the initial guess x, the tolerance θ, and an initial guess for the set $\tilde{S}$. The assertions[1] in Fig. 3.1 suggest that the algorithm will only perform properly if $\tilde{S}$ is indeed a superset of $S(\theta, x)$, that is, if $\tilde{S}$ is an active set. We state the basic characteristic of this algorithm as a lemma.

LEMMA 3.1.2. *If the* sequential adaptive relaxation algorithm *of Fig.* 3.1 *is started with input x and an active set $\tilde{S}$, then the relation $\tilde{S}(\theta, x) \supset S(\theta, x)$ is a loop invariant. On termination, the active set is empty, implying that no residuals beyond the critical tolerance θ remain.*

REMARK 3.1.9. *Termination of the sequential adaptive relaxation algorithm is guaranteed by the positive definiteness of A.*

The following lemma gives a complexity estimate for the sequential adaptive relaxation, which shows that the work bound is directly proportional to the gain in energy (see also Rüde [91]).

LEMMA 3.1.3. *Assume that the* sequential adaptive relaxation *algorithm in Fig.* 3.1 *is started with initial guess x, tolerance θ, and extended active set $\tilde{S} \supset S(\theta, x)$, and let x' denote its result. Then the number of passes W through*

[1]Assertions are used to implement preconditions and postconditions, and to document loop invariants. They are important for proving program correctness and, in practice, as a debugging device.

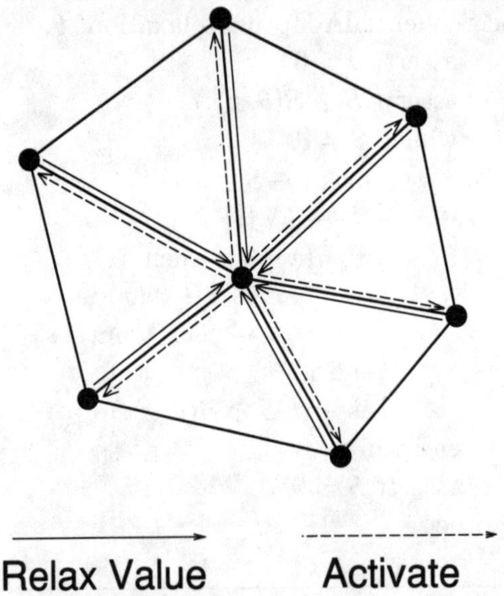

FIG. 3.2. *Elementary relaxation at one node.*

the main loop (and thus the complexity) is bounded by

$$W \le |\tilde{S}| + \left(\|x - x^*\|_E^2 - \|x' - x^*\|_E^2 \right) \frac{\bar{C}}{\theta^2 a_{\min}}, \tag{3.17}$$

where $\bar{C}$ *is the maximal number of connections defined in equation* (3.5), *and*

$$a_{\min} = \min_{i=1,\dots,n} a_{i,i}$$

is the minimal entry in the matrix diagonal. The number of operations is bounded by $W\bar{C}$.

Proof. (See also Proposition 3.2 in [91].) According to Lemma 3.1.1, each pass through the loop that changes x reduces the energy by at least $\theta^2 a_{i,i}$. Note that A positive definite implies $a_{i,i} > 0$. The number of points *changed* is therefore smaller than

$$\gamma = \frac{(\|x - x^*\|_E^2 - \|x' - x^*\|_E^2)}{\theta^2 a_{\min}}.$$

When node i is changed, at most $|\mathrm{Conn}(i)|$ nodes are added to $\tilde{S}$. Thus, in addition to the initial active nodes ($|\tilde{S}|$), at most $\gamma\bar{C}$ nodes can ever become active. This yields the bound given in (3.17).

3.1.2. Simultaneous adaptive relaxation. An active set strategy can also be used in a *simultaneous* (Jacobi-style) relaxation; see Figs. 3.3 and 3.4. The simultaneous relaxation generates symmetric operators that may be useful if the algorithm is to be combined with other components that depend on

```
proc AdaptiveJacobi( ϑ, x, S̃ )
        assert( ϑ > 0 )
        assert( S̃ ⊃ S(ϑ, x) )
        S = {i ∈ S̃| θ_i(x) > θ}
        x ← x + ω ∑_{i∈S} θ_i e_i
        S̃ ← Conn(S)
        assert( S̃ ⊃ S(ϑ, x) )
end proc
```

FIG. 3.3. *Adaptive Jacobi relaxation.*

```
proc SimultaneousAdaptiveRelaxation( θ, x, S̃ )
    assert( θ > 0 )
    assert( S̃ ⊃ S(θ, x) )
    while( S̃ ≠ ∅ )
            AdaptiveJacobi( θ, x, S̃ )
    end while
    assert( S̃ = S(θ, x) = ∅ )
end proc
```

FIG. 3.4. *Simultaneous adaptive relaxation.*

symmetry, like the conjugate gradient method. In the simultaneous variant, the active set $S(\theta, x)$ is explicitly calculated. The main loop in Fig. 3.4 performs a complete sweep of Jacobi with relaxation parameter ω for the equations with large residuals. If S is a set of indices, Conn(S) is defined canonically as an extension of (3.4):

$$\mathrm{Conn}(S) = \bigcup_{i \in S} \mathrm{Conn}(i).$$

The assertion $\tilde{S} \supset S(\theta, x)$ is again a loop invariant.

Classical Jacobi under-relaxation with parameter ω, $0 < \omega \leq 1$, is described by

$$x \longleftarrow x + \omega D^{-1}(b - Ax), \tag{3.18}$$

where $D = \mathrm{diag}(A)$ is the diagonal part of A. The adaptive Jacobi algorithm of Fig. 3.3 can be considered as a perturbation of conventional Jacobi relaxation. This is expressed in the following lemma.

LEMMA 3.1.4. *Assume that the* adaptive Jacobi *routine of Fig.* 3.3 *is started with the initial guess* x, *the tolerance* ϑ, *and an active set that satisfies*

$$\tilde{S} \supset S(\vartheta, x). \tag{3.19}$$

Denote its result by x'. *Then*

$$x' = x + \omega D^{-1}(b - Ax) + \mu, \tag{3.20}$$

where the perturbation is bounded by

$$\|\mu\|_\infty \leq \omega\vartheta. \tag{3.21}$$

Furthermore, at termination the new active set also satisfies (3.19).

Proof. By the definition of $\theta_i(x)$, we have

$$D^{-1}(b - Ax) = \sum_{i=1}^{n} \theta_i e_i,$$

and, correspondingly,

$$\mu = -\omega \sum_{i \notin S} \theta_i e_i.$$

Therefore,

$$\|\mu\|_\infty = \omega \max_{i \notin S} |\theta_i| \leq \omega\vartheta,$$

which proves (3.21). The relation

$$S(\vartheta, x') \subset \text{Conn}(S(\vartheta, x))$$

shows that the new active set will satisfy equation (3.19).

REMARK 3.1.10. *Lemma* 3.1.4 *states that the adaptive Jacobi routine for an appropriately selected critical tolerance converges almost as well as regular Jacobi, which is significant because it will be much cheaper, when* $|\tilde{S}| \ll n$. *In other words, relaxation is only effective where the residuals are large. Most of the computational work for equations with small residuals is wasted.*

The adaptive Jacobi routine can be performed in a loop that is repeated until all residuals are small, similar to the sequential adaptive relaxation of Fig. 3.1. Thus, at termination, the sequential and simultaneous adaptive relaxation will produce results of the same quality. An interesting property of the simultaneous adaptive relaxation scheme is that its error propagation matrix is a symmetric operator.

Conventional sequential relaxation can be easily symmetrized by performing additional sweeps in reverse ordering of the unknowns. For the sequential adaptive relaxation an analogous modification is difficult and may lead to a loss in efficiency. Therefore, algorithms depending on symmetric operators (like preconditioned conjugate gradient iterations) cannot be easily designed on the basis of the sequential adaptive relaxation. For such applications we therefore propose to use the simultaneous adaptive relaxation, as introduced in this section.

3.1.3. Evaluation. This section will provide a limited evaluation of the sequential and simultaneous adaptive relaxation strategies. Lemma 3.1.3 shows that the computational work for the sequential scheme is directly determined by the tolerance parameter θ. This raises the question of which values of θ are suitable. We prepare to answer this question by introducing another lemma that is often useful in the analysis of iterative methods. It relates the current error to the gain in one step of the iteration (see Rüde [83]).

LEMMA 3.1.5. *Consider the linear iteration scheme*

$$x^{(k+1)} = Lx^{(k)} + b,$$

$k = 1, 2, 3, \ldots$, *with the norm*[2] *of* L *bounded by* $\|L\| \le c < 1$, *and assume* x^* *is its fixed point,*

$$x^* = (I - L)^{-1} b.$$

Then

$$\|x^* - x^{(0)}\| \le \frac{1}{1-c} \|x^{(1)} - x^{(0)}\|.$$

Proof.

$$\begin{aligned}
\|x^* - x^{(0)}\| &= \left\| \sum_{k=1}^{\infty} (x^{(k)} - x^{(k-1)}) \right\| \\
&\le \sum_{k=1}^{\infty} \|x^{(k)} - x^{(k-1)}\| \\
&\le \|x^{(1)} - x^{(0)}\| \sum_{k=0}^{\infty} \|L^k\| \\
&\le \|x^{(1)} - x^{(0)}\| \sum_{k=0}^{\infty} c^k \\
&\le \frac{1}{1-c} \|x^{(1)} - x^{(0)}\|.
\end{aligned}$$

We can apply Lemma 3.1.5 to our adaptive algorithm in the following form.

THEOREM 3.1.1. *Let* $D = \operatorname{diag}(A)$ *be the diagonal part of* A *and assume*

$$\|I - D^{-1}A\|_2 \le c < 1. \tag{3.22}$$

Assume that x' *is the result of the* sequential *or the* simultaneous adaptive *relaxation algorithm started with input parameter* θ *and* $\tilde{S} \supset S(\theta, x)$. *Then*

$$\|x' - x^*\|_2 \le \frac{\theta}{1-c}. \tag{3.23}$$

[2]For any vector norm $\|\cdot\|$, the corresponding matrix norm is defined by

$$\|L\| = \sup_{x \ne 0} \frac{\|Lx\|}{\|x\|}.$$

Proof. On termination, $\theta(x') \le \theta$ for all i. Therefore, letting

$$\bar{r} = D^{-1}(b - Ax'),$$

we get $\|\bar{r}\|_2 \le \theta$. Now consider the iterative process

$$y^{(k+1)} = Ly^{(k)},$$

for $k = 0, 1, 2, \dots$, $L = I - D^{-1}A$, and the initial vector $y^{(0)} = x' - x^*$. Thus $\bar{r} = -D^{-1}Ay^{(0)}$, and the first step of the iteration becomes

$$y^{(1)} = Ly^{(0)} = y^{(0)} + \bar{r} = y^{(0)} - D^{-1}Ay^{(0)}.$$

This is just Jacobi's iteration applied to the equation $Ay = 0$ with initial value $y^{(0)} = x' - x^*$. Lemma 3.1.5 then implies

$$||x' - x^*||_2 = ||y^{(0)} - 0||_2 \le \frac{1}{1-c}||y^{(1)} - y^{(0)}||_2 = \frac{||\bar{r}||_2}{1-c} \le \frac{\theta}{1-c}.$$

REMARK 3.1.11. *Unfortunately, the error estimate* (3.23) *involves the operator norm estimate* (3.22). *For the typical case of discretized second order equations,*

$$\|I - D^{-1}A\| = 1 - O(h^2),$$

where h is the mesh size, so we would have $1/(1-c) = O(h^{-2})$. This confirms the observation that classical iterative methods reduce the residual norm $\|r\|$ and the scaled residual norm $\|\bar{r}\|$ quickly, while the error norm $\|x^* - x\|$ may stay large. This is the effect of the ill-conditioning of the system matrix A. As for any basic iterative method, the performance of the adaptive relaxation scheme deteriorates with increased system size. In this respect, the adaptive relaxation scheme is no better than the straightforward Gauß–Seidel relaxation scheme. As we will see further below, this changes when the multilevel structure gets involved.

REMARK 3.1.12. *If the algorithm in Fig.* 3.1 *or* 3.4 *is applied to the discrete systems arising in standard situations, like Poisson's equation in a plane domain discretized by the* 5*-point stencil, θ must be chosen as $O(h^2\epsilon)$, where ϵ is the desired accuracy and h is the mesh size. From Lemma* 3.1.3, *we then obtain a computational complexity for sequential adaptive relaxation of $O(h^{-4}\epsilon^{-1})$.*

Thus, the adaptive versions of the classical relaxation methods have the same order of complexity as straightforward Gauß–Seidel or Jacobi iteration when they are used as solvers. Of course, some improvement could be obtained by dynamically adapting θ, starting with large values and slowly reducing them to $O(h^2\epsilon)$. However, this would not reduce the *order* of the computational work. The adaptive relaxation would only have an advantage at the beginning of the iteration, when for some equations relaxation is more efficient than for others.

If the residuals are relaxed in the order of their size, we will quickly reduce them all to about the same size, so that the active set strategy will not result in anything much different from classical relaxation.

Thus, adaptive relaxation by itself is not competitive with efficient iterative methods for standard applications. Instead, adaptive relaxation should be perceived as a device to augment an iterative method, when some nonuniformity makes it desirable to perform the operations only *locally*. The following section is devoted to extending the adaptive relaxation idea to the multilevel case, where the concept provides an essential improvement over nonadaptive iterations.

3.2. Algebraic structure of multilevel algorithms

In the multilevel context, relaxation is only used as a smoother, not as a solver. A smoother need not eliminate the error but only make it well approximatable by a system with fewer unknowns. For discretized PDEs, this means that the error must be so smooth that it can be represented on a coarser grid; from the algebraic point of view, this means that the error must lie (approximately) in the range of the interpolation operator.

To keep the presentation simple, we will restrict the discussion to *variational* algorithms where the coarse level equations are derived by the so-called Galerkin condition. We assume that we are given a sequence of spaces

$$\{\mathbb{R}^{n_k}\}_{k=0,\ldots,K}, \tag{3.24}$$

with $n_{k+1} \geq n_k$, $k = 0, \ldots, K-1$, and with $n_K = n$ as well as *projections*

$$I_{k+1}^{k} : \mathbb{R}^{n_{k+1}} \longrightarrow \mathbb{R}^{n_k}$$

that are available for $k = 0, \ldots, K-1$. We define the product projections by

$$I_K^k = I_{k+1}^k I_{k+2}^{k+1} \cdots I_K^{K-1}, \tag{3.25}$$

which maps $\mathbb{R}^n$ to any of its subspaces $\mathbb{R}^{n_k}$, $k = 0, 1, \ldots, K-1$. The transposed operators

$$I_k^{k+1} = (I_{k+1}^k)^T \quad \text{and} \quad I_k^K = (I_K^k)^T \tag{3.26}$$

are the corresponding *interpolation operators.*

The coarse spaces $\mathbb{R}^{n_k}$, $k < K$, can now be used to *approximate* the original system. The equation

$$A_k x_k = b_k, \tag{3.27}$$

with $A_k = I_K^k A I_k^K$ and $b_k = I_K^k b$, is the *projection* of (3.1) onto the subspace $\mathbb{R}^{n_k}$. The sequence of coarse grid systems (3.27) may now be used to accelerate the convergence of iterative solvers for (3.1).

A vector $x \in \mathbb{R}^n$ can be represented *nonuniquely* as

$$x = \sum_{k=0}^{K} I_k^K x_k = \hat{I}^T \hat{x},$$

where

$$\hat{I}^T = [I_K^K I_{K-1}^K \cdots I_1^K I_0^K],$$

$I_K^K = \mathrm{Id}$ is the identity, and $x_k \in \mathbb{R}^{n_k}$ for $k = 1, \ldots, K$ (see Griebel [41]). This representation is closely related to the hierarchical basis concept; see §3.6. With the nonunique representation, the minimization problem (3.3) becomes

$$\min_{\substack{x_k \in \mathbb{R}^{n_k} \\ k=0,\ldots,K}} \left(\left(\sum_{k=0}^{K} I_k^K x_k \right)^T A \left(\sum_{k=0}^{K} I_k^K x_k \right) - 2 \left(\sum_{k=0}^{K} I_k^K x_k \right)^T b \right). \tag{3.28}$$

Forming the normal equations for the minimization problem (3.28) with a vector of variables

$$\hat{x} = \begin{bmatrix} x_0 \\ \vdots \\ x_K \end{bmatrix} \in \mathbb{R}^{\hat{n}},$$

where $\hat{n} = \sum_{\nu=0}^{K} n_\nu$, we obtain a *singular, positive semidefinite* system of equations

$$\hat{A}\hat{x} = \hat{b}, \tag{3.29}$$

where

$$\hat{A} = \hat{I} A \hat{I}^T = \begin{bmatrix} A_0 & I_1^0 A_1 & & \cdots & I_K^0 A_K \\ A_1 I_0^1 & A_1 & I_2^1 A_2 & \cdots & I_K^1 A_K \\ \vdots & & & & \vdots \\ A_K I_0^K & & \cdots & A_K I_{K-1}^K & A_K \end{bmatrix} \tag{3.30}$$

and

$$\hat{b} = \hat{I} b = \begin{bmatrix} I_K^0 b \\ I_K^1 b \\ \vdots \\ b \end{bmatrix}.$$

Though (3.29) is singular, it is solvable because $\hat{b} \in \mathrm{Image}(\hat{A})$. From any solution $\hat{x}$ of (3.29), we get the *unique* solution of (3.1) by forming the sum

$$x = \sum_{k=0}^{K} I_k^K x_k = \hat{I}^T \hat{x}.$$

A usual multigrid V-cycle with a single sweep of presmoothing and no postsmoothing can be interpreted as a relaxation of the semidefinite quadratic minimization problem (3.28) or its gradient equation (3.29). Other (variational) multigrid algorithms can be simulated by different relaxation orderings.

A number of iterative algorithms, like Gauß–Seidel relaxation in any ordering, or like steepest descent and conjugate gradient iteration (see Luenberger [53]), do not depend on the scaling of the system to be solved. Such self-scaling algorithms converge, despite the fact that the system (3.29) is singular. Non-self-scaling methods, like Jacobi iteration, are generally divergent when applied to (3.29). These iterations must be modified by either adding an under-relaxation parameter (the optimal parameter leads to the steepest descent method) or by treating the different levels sequentially.

Alternatively and equivalently to the nonunique representation, we can apply all changes directly to the finest mesh and use the finest mesh representation $x = \sum_{k=0}^{K} I_k^K x_k$ of the solution exclusively. This leads to a representation equivalent to *Unigrid*; see McCormick and Ruge [61] and [58].

For the semidefinite matrix $\hat{A}$ with eigenvalues

$$0 = \lambda_1 = \cdots = \lambda_{k_0 - 1} < \lambda_{k_0} \leq \lambda_{k_0+1} \leq \cdots \leq \lambda_{\hat{n}},$$

we can introduce the *generalized condition number*

$$\kappa(\hat{A}) = \frac{\lambda_{\hat{n}}}{\lambda_{k_0}}.$$

Next we introduce the *discrete additive Schwarz* operator for the sequence of spaces (3.24) with respect to the projections and interpolations (3.25), (3.26) in analogy to Definition 2.1.4:

$$P_{\mathbb{R}^n} = \sum_{k=0}^{K} I_k^K S_k I_K^k A = \hat{I}^T \hat{S} \hat{I} A, \tag{3.31}$$

where S_k is an operator corresponding to the bilinear forms b_k, defined in (2.13) and

$$\hat{S} = \begin{bmatrix} S_0 & 0 & & \cdots & 0 \\ 0 & S_1 & 0 & \cdots & 0 \\ \vdots & & & & \vdots \\ 0 & \cdots & 0 & S_{K-1} & 0 \\ 0 & & \cdots & 0 & S_K \end{bmatrix}.$$

LEMMA 3.2.1. *All nonzero eigenvalues of the* $\hat{S}$-preconditioned *semidefinite matrix*

$$\hat{S}\hat{A} = \hat{S}\hat{I}A\hat{I}^T \tag{3.32}$$

are also eigenvalues of the additive Schwarz operator $\hat{I}^T\hat{S}\hat{I}A$. *If* $\hat{I}$ *has full rank then all eigenvalues of the additive Schwarz operator are also eigenvalues of the preconditioned semidefinite matrix, and the generalized condition number of* $\hat{S}\hat{A}$ *is the same as the condition number of the additive Schwarz operator.*

Proof. (See Griebel [41].) For any eigenvector $y \in \mathbb{R}^{\hat{n}}$ with

$$\hat{S}\hat{I}A\hat{I}^T y = \lambda y,$$

the vector $x = \hat{I}^T y \in \mathbb{R}^{\hat{n}}$ satisfies

$$\hat{I}^T \hat{S} \hat{I} A x = \lambda x.$$

Thus, either $x = \hat{I}^T y = 0$ such that $y \in \text{kernel}(\hat{A})$ and $\lambda = 0$, or x is an eigenvector with eigenvalue λ of $\hat{I}^T \hat{S} \hat{I} A$.

Assume that $\hat{I}^T$ has full rank and that $y_1, \ldots, y_{\hat{n}}$ are the $\hat{n}$ eigenvectors of $\hat{S}\hat{A}$ such that

$$\hat{S}\hat{A} y_i = \lambda_i y_i.$$

Without loss of generality we can assume that $\hat{I}^T y_i \neq 0$ for $1 \leq i \leq n$. Thus we have found a full system of eigenvectors $x_i = \hat{I}^T y_i$ of the additive Schwarz operator $\hat{I}^T \hat{S} \hat{I} A$. Each such eigenvalue corresponds uniquely to an eigenvalue of the semidefinite matrix $\hat{S}\hat{I}\hat{A}\hat{I}^T$. Therefore, $\hat{I}^T \hat{S} \hat{I} A$ and $\hat{S}\hat{A}$ have the same generalized condition number.

REMARK 3.2.1. *With Lemma* 3.2.1, *the conjugate gradient method in the system* (3.29), *preconditioned by* $\hat{S}$, *can be shown to be algebraically equivalent to the BPX method.*

REMARK 3.2.2. *Note that the nonunique representation and Unigrid are mainly useful as methods to* simulate and study *different multilevel algorithms. Both methods are too inefficient to be of direct practical value. They are, however, useful for analyzing multilevel algorithms that must then be implemented in a conventional way.*

Classical multilevel algorithms fix their attention on *one* level at a time. As much information as possible is collected in one x_k. This x_k is then updated, keeping all others fixed, resulting in an equation for x_k of the form

$$\underbrace{I_K^k A I_k^K}_{A_k} x_k = I_K^k \left(b - A \sum_{\nu=0,\ \nu\neq k}^{K} I_\nu^K x_\nu \right) \tag{3.33}$$

for $k = 0, 1, \ldots, K$. These systems are usually relaxed consecutively, as in the algorithm in Fig. 3.5, where procedure *relax* can be any of several iterative schemes, including Jacobi or Gauß–Seidel.

We consider the special case of Jacobi relaxation for equation (3.33) for a given level k, which is defined by

$$x_k \longleftarrow x_k + \omega D_k^{-1} I_K^k \left(b - A \sum_{\nu=0}^{K} I_\nu^K x_\nu \right), \tag{3.34}$$

where D_k is the diagonal part of $A_k = I_K^k A I_k^K$ and ω is an appropriate (under-)relaxation parameter. Below we will study adaptive relaxation as an elementary process for each level.

LEMMA 3.2.2. *Let x be the input of algorithm* Simple-V-Cycle *of Fig.* 3.5, *and let x' be its output. The error after application of* Simple-V-Cycle *is given by*

$$\begin{aligned}(x' - x^*) = {} & (I - \omega I_0^K D_0^{-1} I_K^0 A)(I - \omega I_1^K D_1^{-1} I_K^1 A) \quad \cdots \\ & \cdots (I - \omega D_K^{-1} A)(x - x^*).\end{aligned} \tag{3.35}$$

```
proc Simple-V-Cycle( x )
        for ν = K to 0 step −1
                relax according to equation (3.34)
        end for
end proc
```

FIG. 3.5. *Simple multigrid V-cycle with only one step of presmoothing.*

Proof. Note that the action of equation (3.34) on any level is independent of the particular representation of $x = \sum_{\nu=0}^{K} I_\nu^K x_\nu$. For the error, the effect of each such step is a multiplication by

$$(I - \omega I_\nu^K D_\nu^{-1} I_K^\nu A),$$

so that the operator product in equation (3.35) is generated by the definition of Simple-V-Cycle in Fig. 3.5.

REMARK 3.2.3. *The representation of a multigrid algorithm in the product form of equation* (3.35) *depends on the use of a* fully variational *setup. This is the basis for the V-cycle analysis in Braess and Hackbusch* [22], *McCormick and Ruge* [60], *and Bramble, Pasciak, Wang, and Xu* [24].

REMARK 3.2.4. *To make the execution efficient, practical multilevel algorithms do not use the nonunique representation or unigrid representation of the approximate solution directly, but propagate all information to the level currently being processed.*

Many multilevel algorithms, including the hierarchical basis and BPX preconditioners (see Yserentant [111] and Bramble, Pasciak, and Xu [25]), have an algebraically equivalent interpretation as iterative methods in the semidefinite system (3.28).

The coarse spaces $\mathbb{R}^{n_0}, \mathbb{R}^{n_1}, \ldots, \mathbb{R}^{n_{K-1}}$ *add more minimization search directions* to the original system. Gauß–Seidel relaxation for the original system will almost always stall, because the effective reduction of smooth errors is impossible along any of the coordinate directions, that is, by any elementary relaxation step. A multilevel method, in contrast, is usually able to reduce the error quickly because, in the enlarged system, broader based directions always provide good error reduction. This is the essence of the multigrid method, as well as the multilevel preconditioning algorithms.

3.3. Application of the theory of multilevel splittings

We will now study how the abstract theoretical results for multilevel algorithms from Chapter 2 can be used to analyze a multilevel adaptive iteration.

THEOREM 3.3.1. *Assume that the matrices A_k, $0 \le k \le K$, $A_K = A$, originate from a discretization of problem* (1.1) *or* (1.2) *by a nested sequence of finite element meshes such that the prerequisites of Theorem* 2.2.2 *are satisfied. In particular, let I_k^{k+1} denote the discrete representations of the associated interpolation operators defined by embedding the spaces. Then there exist* *constants* $0 < c_1 \leq c_2 < \infty$ *independent of* K *such that, for all* $x \in \mathbb{R}^n$, *we have*

$$c_1 x^T Ax \leq x^T A \left(\sum_{\nu=0}^{K} I_k^K D_k^{-1} I_K^k \right) Ax \leq c_2 x^T Ax, \tag{3.36}$$

where $D_k = \text{diag}(A_k)$ *is the diagonal part of* A_k *for* $k = 1, 2, 3, \ldots$ *and* $D_0 = A_0$.

Proof. In the proof we must relate the algebraic language of this chapter to the finite element theory in Chapter 2. To this end we identify a finite element function $u_k \in V_k$ with the vector of its nodal values $\vec{u}_k \in \mathbb{R}^{n_k}$. Then

$$a(u_k, u_k) = \vec{u}_k^T A_k \vec{u}_k,$$

and

$$\|\vec{u}_k\| \stackrel{\text{def}}{=} \sqrt{\vec{u}_k^T A_k \vec{u}_k}$$

defines a norm equivalent to $\|u_k\|_{H^1}$. Furthermore, we define

$$b_k(u_k, u_k) = \vec{u}_k^T D_k^{-1} \vec{u}_k$$

for $\vec{u}_k \in \mathbb{R}^{n_k}$. Clearly, the bilinear form $b_k(\cdot, \cdot)$ is equivalent to $2^{2k}(u_k, u_k)_{L_2}$, that is, there exist constants $0 < c_5 \leq c_6 < \infty$ independent of k such that

$$c_5 2^{2k}(u_k, u_k)_{L_2} \leq b_k(u_k, u_k) \leq c_6 2^{2k}(u_k, u_k)_{L_2},$$

$k = 1, 2, 3, \ldots$. The inner product $b_0(\cdot, \cdot)$ is equivalent to the H^1-inner product. We can now link this setup to the abstract results in Chapter 2. In particular, Theorem 2.2.2 asserts that $\{\mathbb{R}^{n_k}\}_{j=0,\ldots,K}$ is a stable splitting of $\mathbb{R}^n$ with stability constant uniformly bounded in K. The additive Schwarz operator in discrete notation is

$$P_{\mathbb{R}^n} = \sum_{k=0}^{K} I_k^K D_k^{-1} I_K^k A; \tag{3.37}$$

see also (3.31). With Theorem 2.1.1, we conclude that there exist constants $0 < c_1 \leq c_2 < \infty$ such that

$$c_1 \leq \lambda_{\min}(P_{\mathbb{R}^n}) \leq \lambda_{\max}(P_{\mathbb{R}^n}) \leq c_2.$$

Written in matrix notation, this is just assertion (3.36).

COROLLARY 3.3.1. *Assume that the assumptions of Theorem* 3.3.1 *hold and that* $S_k^{-1} = D_k = \text{diag}(A_k)$ *for* $k = 1, 2, \ldots, K$ *and* $S_0^{-1} = D_0 = A_0$. *The condition number of the discrete additive Schwarz operator* $P_{\mathbb{R}^n}$ *of* (3.31), *which is the same as the generalized condition number of the preconditioned semidefinite matrix* $\hat{S}\hat{A}$ *of* (3.32), *is bounded by* c_2/c_1.

Proof. With $v = A^{1/2}x$, Theorem 3.3.1 implies

$$c_1 \leq \frac{v^T A^{1/2} \left(\sum_{k=0}^{K} I_k^K D_k^{-1} I_K^k \right) A^{1/2} v}{v^T v} \leq c_2,$$

for some constants $0 < c_1 \leq c_2 < \infty$ independent of K and for all $v \in \mathbb{R}^n$. Thus all the eigenvalues of the preconditioned matrix satisfy

$$c_1 \leq \lambda\left(\sum_{k=0}^{K} I_k^K D_k^{-1} I_K^k A\right) \leq c_2. \tag{3.38}$$

This shows the condition number estimate.

We now define the discrete analogue of the *residual* of Definition 2.6.1.

DEFINITION 3.3.1. *For any level* $k = 0, 1, \ldots, K$ *we define the* **current residual** *of* $x \in \mathbb{R}^n$ *(see also Definition* 3.1.4*) by*

$$r_k = I_K^k(b - Ax), \tag{3.39}$$

and the **current scaled residual** *by*

$$\bar{r}_k = D_k^{-1} r_k = D_k^{-1} I_K^k (b - Ax). \tag{3.40}$$

The current scaled residual on level K is

$$\bar{r}_K = \bar{r} = \sum_{i=1}^{n_K} \theta_i e_i.$$

On the coarser levels it is obtained by restriction of the fine grid residual.

COROLLARY 3.3.2. *Under the assumptions of Theorem* 3.3.1 *there exist constants* $0 < c_1 \leq c_2 < \infty$ *such that the Euclidean norms are bounded by*

$$\frac{1}{c_1}\left\|\sum_{k=0}^{K} I_k^K \bar{r}_k\right\|_2 \leq \|x - x^*\|_2 \leq \frac{1}{c_2}\left\|\sum_{k=1}^{K} I_k^K \bar{r}_k\right\|_2. \tag{3.41}$$

Proof. This follows from the observation that

$$x - x^* = \left(\sum_{k=0}^{K} I_k^K D_k^{-1} I_K^k A\right)^{-1}\left(\sum_{k=0}^{K} I_k^K \bar{r}_k\right),$$

and the eigenvalue estimates (3.38) show equation (3.41).

REMARK 3.3.1. *The error estimate* (3.41) *is the discrete analogue of the abstract estimate in Theorem* 2.6.1.

REMARK 3.3.2. *Corollary* 3.3.2 *could also be proved using Lemma* 3.1.5 *applied to a so-called BPX iteration. A steepest descent for the preconditioned system*

$$\sum_{k=0}^{K} I_k^K D_k^{-1} I_K^k A$$

can be modeled as a damped Jacobi iteration, so that Lemma 3.1.5 *applies. Thus the upper bound of equation* (3.41) *is a consequence of the* $O(1)$ *convergence rate of the BPX steepest descent iteration, which in turn is a consequence of the condition number estimate in Corollary* 3.3.1.

Besides the L_2-error, we can also bound the energy norm.

COROLLARY 3.3.3. *Under the hypothesis of Theorem* 3.3.1,

$$\frac{1}{c_2}\sum_{k=0}^{K}\bar{r}_k^T D_k \bar{r}_k \le (x-x^*)^T A(x-x^*) \le \frac{1}{c_1}\sum_{k=0}^{K}\bar{r}_k^T D_k \bar{r}_k \tag{3.42}$$

and

$$\frac{1}{c_2}\sum_{k=0}^{K}{r_k}^T D_k^{-1} r_k \le (x-x^*)^T A(x-x^*) \le \frac{1}{c_1}\sum_{k=0}^{K}{r_k}^T D_k^{-1} r_k. \tag{3.43}$$

Proof. Both estimates follow from equation (3.36) applied to the error $x-x^*$ using the respective equations

$$\bar{r}_k = -D_k^{-1} I_K^k A(x-x^*)$$

and

$$r_k = -I_K^k A(x-x^*).$$

REMARK 3.3.3. *Corollary* 3.3.3 *is the discrete analogue of Theorem* 2.6.2.

The essence of Corollaries 3.3.2 and 3.3.3 can be formulated as:

> *If the scaled residuals are small* **on all levels**, *then the error must be small.*

REMARK 3.3.4. *Corollaries* 3.3.2 *and* 3.3.3 *can clearly serve as error estimates for a given approximation. This is a* static *interpretation, where the corollaries are used to make statements about a fixed situation. There is also a* dynamic *interpretation of these corollaries. If elementary relaxation steps* (3.6) *are used, then Lemma* 3.1.1 *shows that the iteration will converge to an acceptable solution if we use* critical tolerances *to restrict the work to residuals that are approximately no smaller than the size of the desired accuracy.*

In contrast to the algorithm on a single level (see also Remarks 3.1.11 and 3.1.12) we need not consider elementary relaxations with small residuals. This in turn makes Lemma 3.1.1 a powerful tool that guarantees fast convergence. Assume that we have a system with K levels and that the error satisfies

$$\|x-x^*\|_E^2 > \epsilon^2.$$

From equation (3.42), we can conclude that there exists an equation with current scaled residual of size at least $O(\epsilon/2^K)$. More precisely, there exist a constant $c>0$ and indices k and i, $0 \le k \le K$ and $1 \le i \le n_k$, such that

$$(e_i^T \bar{r}_k)^2 \ge c\frac{\epsilon^2}{4^K}.$$

If equation i on level k is relaxed, Lemma 3.1.1 guarantees a reduction in the square of the energy norm proportional to

$$c\frac{\epsilon^2}{4^K}.$$

By performing $O(4^K)$ such elementary relaxation steps, the error norm is reduced by an amount proportional to ϵ and independent of K. Thus, with work proportional to the size of the problem, we get convergence, independent of the number of levels. This of course requires that we restrict the relaxation to equations with residuals of size at least $O(\epsilon/2^K)$, as can be accomplished by the adaptive relaxation algorithms.

It is important to note that the same argument holds for locally refined meshes, provided the refinement is appropriate in the sense that the (hypothetical) residuals of the solution interpolated to the *global* finest grid are small in the regions that have not been refined. Only in the refined region may we have large residuals that can be eliminated by relaxation. This is precisely why the refinement in this region is necessary. For either globally or locally refined meshes the theory here assures accuracy of ϵ in the norm $\|\cdot\|_{L_2}$ only when relaxation has reduced the scaled residuals to satisfy $\|\bar{r}_k\|_\infty \leq \epsilon_k$, where

$$\sum_{k=1}^{K} \epsilon_k = O(\epsilon). \tag{3.44}$$

Experimental results, however, suggest that equation (3.44) is conservative, and it seems sufficient to use $\epsilon_k = \epsilon$, which would only strictly guarantee an L_2-error estimate of $\sum_{k=1}^{K} \epsilon_k = K \cdot \epsilon$.

The above discussion clearly shows what the role of relaxation in the multilevel context should be:

> *The residuals on all levels must be reduced approximately to the size of the desired accuracy.*

The adaptive relaxation technique satisfies these requirements exactly. By its construction, it additionally provides suitable stopping criteria, and it delivers diagnostic information. If used within a multilevel algorithm, adaptive relaxation is efficient because, according to Lemma 3.1.1, each (executed) relaxation step yields a substantial reduction of the error.

REMARK 3.3.5. *Corollaries* 3.3.2 *and* 3.3.3 *can be used directly to guide the* adaptive *mesh refinement process. They apply to both* algebraic *error and* discretization *error. The bounds are uniform in* K *and can thus be applied for* $K \to \infty$. *In the limit case they become true bounds for the discretization error. Clearly, this unified interpretation of errors is of high practical relevance because it provides powerful criteria to guide the interplay between iteration and refinement. This is exploited in the* virtual global grid *refinement technique; see* §§3.7 *and* 4.4.8. *Note that the residual for each unknown* i *on each level* k *is a measure of how much the error with respect to the continuous solution can be improved by the corresponding relaxation.*

This leads naturally to the interpretation of a finite element problem as one being embedded in an infinite sequence of refined meshes. At any time the solution is defined on all levels by interpolation from the coarser ones. Thus there is no real distinction between algebraic and discretization errors because they are interpreted to have the same source and they are decreased by the same means: relaxation of the unknowns with large residuals. This is the key idea behind the *virtual global grid* refinement technique discussed in §§3.7 and 4.4.8.

Summing up, the multilevel concept and the statements derived in this section help to solve three of the crucial points in the numerical treatment of elliptic equations: They provide *fast solvers*, *error estimates*, and a direct means to handle *adaptivity.*

3.4. Multilevel adaptive iteration

3.4.1. Tracing dependencies between levels. We now turn to the study of practical implementation of multilevel adaptive algorithms. Unfortunately, the most straightforward idea of using adaptive relaxation directly for the additive Schwarz operator matrix (3.31) or the semidefinite matrix (3.29) is not yet satisfactory. This is for the same reason that multilevel algorithms usually do not work with this matrix explicitly: The preconditioned matrix is not sparse in the sense of Remark 3.1.2, and it must be used in the *factored form* of operator (3.31), where the levels are treated consecutively and independently. The main point here is to minimize the number of expensive inter-level transfers.

This can, for example, be implemented in algorithmic structures that are similar to the Simple-V-Cycle in Fig. 3.5, where the generic relaxation is replaced, perhaps, by the sequential adaptive relaxation algorithm of Fig. 3.1. Of course, this raises the issue of how the extended active sets should be initialized on each level.

DEFINITION 3.4.1. *The active set for level k is defined by*

$$S_k(\vartheta_k, x) = \left\{ i \mid \ 1 \le i \le n_k, \ |e_i^T \bar{r}_k| > \vartheta_k \right\}, \tag{3.45}$$

where $\bar{r}_k$ *is the current scaled residual of level k, as defined in equation* (3.39).

The trivial construction of an initial superset $\tilde{S}_k$ of $S_k(\vartheta_k, x)$ by

$$\tilde{S}_k = \{1, 2, \ldots, n_k\}$$

is possible, but this neglects useful information. To exploit the locality of operations, we need to transfer information between levels. This is done as if adaptive relaxation were applied to the semidefinite system (3.28) directly. However, inter-level information is collected and only transferred when the algorithm switches between levels. How this is arranged in practice depends on the cycling strategy.

The adaptive relaxation scheme must be augmented to monitor all changes to the current solution on the present level. This information is collected and must be passed to the neighboring levels when required. To give a concise description, we introduce the following notation.

DEFINITION 3.4.2. *For a given matrix*

$$M = (\mu_{i,j})_{1 \le i \le m_1, 1 \le j \le m_2} \in \mathbb{R}^{m_1} \times \mathbb{R}^{m_2}$$

and set $S \subset \{1, 2, \ldots, m_2\}$ *of indices, let* MS *denote the set of indices defined by*

$$MS = \{i | \, 1 \leq i \leq m_1, \, \exists j \in S : \mu_{i,j} \neq 0\}.$$

REMARK 3.4.1. *If sets are identified with Boolean vectors,* M *denotes a Boolean matrix multiplication. In this notation the mapping* "Conn," *as defined in equation* (3.4), *could be written as*

$$\text{Conn} = A - D.$$

REMARK 3.4.2. *Note that, for two matrices* A *and* B, *we only have*

$$(AB)S \subset A(BS),$$

not necessarily equality.

In Fig. 3.6, we display a generic multilevel adaptive iteration routine. The inner loop is precisely the adaptive relaxation scheme, applied on the different levels k. It is augmented with sets U_k and V_k to provide information on where the relaxation causes changes on higher or lower levels. These sets can also be used to restrict the operations performed for a projection or interpolation to specific subsets of the domain. This will be discussed in §3.6.2.

Whether "upward" or "downward" is selected depends on the cycling strategy. In either direction, the information accumulated in the set U_k or V_k is exploited to initialize the active set on the new level. Furthermore, the new U-set or V-set on the new level is updated so that information will be passed successively through all levels.

3.4.2. Cycling strategies. The cycling strategy is not yet specified in the algorithm in Fig. 3.6. It is thus still a *nondeterministic* algorithm. There are many possible variants, including "classical" ones, like V- or W-cycles, or *accommodative* cycles (where the switch between levels depends on information accumulated within the algorithm); see Brandt [27].

Clearly, when all dynamic residuals on a given level have become *small*, the algorithm must either proceed to another level or terminate. In a V-cycle or W-cycle the next level is determined a priori. For these algorithms a cycle will generally terminate with the residuals on the last level eliminated, but large residuals may be present on other levels. This information can be used to decide whether the cycle must be repeated to obtain an acceptable solution. A further cycle with the same tolerances will now only touch unknowns in the active sets, and will often be significantly cheaper than the previous cycles, because typically only few unknowns have failed to converge in a single cycle.

For this idea to work, the adaptive relaxation strategy need not completely eliminate all residuals on each level. In a sequential relaxation we may think of limiting the number of elementary relaxations to be proportional to the number of unknowns on that level, before we force the method to go to the next level. This might be useful because it makes the algorithm less sensitive

```
proc MLAI( x, S̃_1, ϑ_1, S̃_2, ϑ_2, ..., S̃_K, ϑ_K )
    assert(S̃_k ⊃ S_k(ϑ_k, x), 1 ≤ k ≤ K)
    U_k ← ∅, V_k ← ∅, for k = 1, 2, ..., K
    k ← 1
    while(⋃_{k=1}^K S̃_k ≠ ∅)
        while(S̃_k ≠ ∅)
            pick i ∈ S̃_k
            S̃_k ← S̃_k \ {i}
            if |e_i^T r̄_k| > ϑ_k then
                x_k ← x_k + (e_i^T r̄_k) e_i
                S̃_k ← S̃_k ∪ Conn_k(i)
                U_k ← U_k ∪ {i}
                V_k ← V_k ∪ {i}
            end if
            assert(S̃_k ⊃ S_k(ϑ_k, x))
        end while
        if "upward" then
            k ← k + 1
            U_k ← U_k ∪ I_{k-1}^k U_{k-1}
            S̃_k ← S̃_k ∪ A_k I_{k-1}^k U_{k-1}
            U_{k-1} ← ∅
        else "downward"
            k ← k - 1
            V_k ← V_k ∪ I_{k+1}^k V_{k+1}
            S̃_k ← S̃_k ∪ I_{k+1}^k A_{k+1} V_{k+1}
            V_{k+1} ← ∅
        end if
        assert(S̃_k ⊃ S_k(ϑ_k, x))
    end while
end proc
```

FIG. 3.6. *Multilevel adaptive iteration (MLAI).*

to the correct selection of the critical tolerances. Otherwise, if the critical tolerances were too small, too many iterations would be performed without changing levels.

Consider the extreme case that the number of relaxation steps on each level is set equal to the number of equations, that the tolerance is selected as 0, and that we choose a fixed standard cycling strategy. The method now degenerates to a classical multigrid algorithm with global smoothing and fixed cycling.

If a fixed cycle is combined with simultaneous iteration, it is also possible to replace the *simultaneous adaptive relaxation* scheme with a fixed number of *adaptive Jacobi* sweeps. This leads to a method with simple structure that lends itself to an interpretation as a *perturbed* classical multigrid method. We will analyze this type of algorithm in §3.5.

Alternatively, the number of elements in the current active sets on the different levels can be used to adaptively select the cycling strategy. In the FAMe algorithm of [91], for example, the algorithm uses the "downward" branch as long as V_k is not empty. Here, a simple induction argument shows that each subproblem $A_k x_k = b_k$ occurring in the process is solved up to the specified precision before returning to finer levels. This algorithm is therefore robust, although its efficiency depends sensitively on the selection of suitable tolerance parameters.

3.4.3. Selection of the critical tolerances. If the ϑ-values are chosen like

$$\vartheta_k = \epsilon 2^{k-K},$$

then it is guaranteed that the L_2-error at termination is bounded by

$$\sum_{k=0}^{K} \epsilon 2^{k-K} = O(\epsilon),$$

independent of the number of levels. Practical experience suggests that it is better to choose all ϑ_k to have the same value even though the error bound (3.41) would then grow with the number of levels. Such an adverse effect on the error has not been observed in practice, suggesting that the current L_2-bounds (containing the *sum* of the residuals over the levels) are not yet sharp.

Generally, for a fixed grid, we must be careful to select the critical tolerances so that they are not too small, because otherwise we will force the algorithm to solve the algebraic system to levels below discretization errors; see the discussion in the previous section. The selection of the ϑ-values should therefore be derived from error estimators.

Beyond this, the unified treatment of algebraic errors and discretization errors suggested in Remark 3.3.5 leads to the possibility of combining refinement and iteration. In this context, the critical tolerances should be selected in accordance with the mesh structure. The obvious framework for this is the *virtual global grid* (see §3.7) technique. Ultimately, it means that for a specific selection of critical tolerances the algorithm finds its own mesh based on the same indicators that are used to guide the adaptive iteration.

Naturally, the algorithm can then be used in a *full multigrid* style (see Brandt [28] and Hackbusch [43]), where the progression is driven by successively reduced critical tolerances. Beyond the usual full multigrid algorithms, this method now automatically generates its own approximation spaces, as required to obtain the prescribed accuracy.

REMARK 3.4.3. *Note that the full multigrid theory may require higher order interpolations for computing the initial approximation on a new level. Depending on the type of equation and the norms used to measure the error, a solution procedure may only be optimal if the first approximation is already close enough to the solution. This is not always achieved by an initialization of the new grid with linearly interpolated values; see Brandt* [28].

Obviously, these techniques open a wide arena of algorithmic possibilities whose full potential only begins to become visible. A comprehensive evaluation will require further research and is beyond the scope of the present monograph. The following sections will nevertheless attempt to clarify some of the relevant aspects.

3.5. Analysis of the V-cycle

This section digresses from the main theme to look at a particular multilevel adaptive algorithm. A simple V-cycle with simultaneous adaptive relaxation can be analyzed based on classical multigrid convergence theory. This possibility has been suggested by Leinen and Yserentant [50]. To make the analysis simpler, the adaptive relaxation scheme on each level k is modified such that it can be interpreted as one or more sweeps of regular, global relaxation on that level, perturbed by quantities of size $O(\vartheta_k)$. We will analyze the special case of the V-cycle in Fig. 3.5 (see also Fig. 3.15), where the *adaptive Jacobi* routine of Fig. 3.3 is used as a smoother. The adaptive Jacobi routine has already been analyzed as a perturbed version of a regular Jacobi smoother in Lemma 3.1.4.

THEOREM 3.5.1. *Let x be the input of routine* Simple-V-Cycle *(see Fig. 3.5) for level K, where the relaxation is one sweep of* adaptive Jacobi. *Let $\tilde{x}$ be its output. Assume that the sets satisfy $\tilde{S}_k \supset S(\vartheta_k, x_k)$ before the relaxation on each level k. Assuming*

$$\|I - \omega I_k^K D_k^{-1} A\| \leq 1, \tag{3.46}$$

$k = 1, 2, \ldots, K$, then the perturbation caused by using adaptive Jacobi *instead of regular Jacobi is bounded by*

$$\|\tilde{x} - x'\| \leq \sum_{k=1}^{K} \vartheta_k, \tag{3.47}$$

where x' denotes the result after a Simple-V-Cycle *with unmodified Jacobi smoothing.*

Proof. The result of the V-cycle with regular Jacobi smoothing is given in equation (3.35). The effect of adaptive Jacobi has been described as a perturbation of regular Jacobi smoothing in Lemma 3.1.4, so we can describe $\tilde{x}$ recursively:

$$\begin{aligned}
\tilde{x}_0 &= x, \\
\tilde{x}_k - x^* &= (I - \omega I_k^K D_k^{-1} A)(\tilde{x}_{k-1} - x^*) + \mu_k, \\
\tilde{x} &= \tilde{x}_K,
\end{aligned}$$

$k = 1, 2, \ldots, K$. This leads directly to the following recursive formula for the accumulated perturbations

$$\begin{aligned} \tilde{\mu}_0 &= 0, \\ \tilde{\mu}_k &= (I - \omega I_k^K D_k^{-1} A)\tilde{\mu}_{k-1} + \mu_k, \\ \tilde{x} - x' &= \tilde{\mu}_K, \end{aligned}$$

$k = 1, 2, \ldots, K$. With assumption (3.46) we have

$$\|\tilde{\mu}_K\| \le \sum_{k=1}^{K} \|\mu_k\| \le \sum_{k=1}^{K} \vartheta_k. \tag{3.48}$$

REMARK 3.5.1. *Theorem* 3.5.1 *analyzes the simple case of a V-cycle with one single sweep of presmoothing and no postsmoothing. Here it suffices to argue that the following corrections within the same cycle cannot enlarge the error; in particular, they cannot enlarge the* μ_k *perturbations, so that their total effect must be bounded by* $\sum_{k=1}^{K} \vartheta_k$. *The generalization to several smoothing sweeps is obvious. For W-cycles, the analysis must be more careful, showing that later corrections will even* reduce *the effect of the perturbations, so that their sum is still strictly bounded.*

In general, we conclude that an adaptive multilevel iteration has the same convergence properties as the corresponding conventional multigrid algorithm, except for some perturbations that can be easily controlled by choosing appropriate tolerance parameters ϑ_k.

Though the analysis outlined above is far from being sharp, it shows why and when multilevel adaptive iteration is superior to classical multigrid. This is always the case when the characteristic features of the problem vary in the solution domain, so that certain subdomains need more attention than others. Typical examples are boundary and interior layers, jumps in the coefficients, singularities, and effects caused by adaptive mesh refinement.

This analysis does not provide quantitative results, because these depend on problem-dependent features that must be properly specified before they can be analyzed.

3.6. Implementation using hierarchical transformations

In this section we give a more detailed description of multilevel adaptive iteration as it is used in the FAMe. We will work out the details of procedure MLAI of Fig. 3.6 based on the calculus of hierarchical transformations. In particular, we describe how the nonunique representation of the solution can be made unique, and how the transfer of information between levels is optimized to increase the efficiency of the algorithm.

The algebraic reformulation of multilevel methods with hierarchical transformations generally has the advantage of leading to an efficient implementation of the *full approximation storage scheme* (FAS; see Brandt [28]). It is also well suited for local refinement; see McCormick and Rüde [59]. To set up the appropriate hierarchical transformations, we assume a nested system of finite element spaces, as described in Chapter 2 or in Rüde [86].

As remarked above, we attempt to exploit the locality of operations. The mesh defines dependencies within a level that are exploited by the adaptive relaxation algorithms. The dependencies between levels are also of local nature. Typically, a single change on one level will affect only few unknowns on the next coarser and next finer level. These dependencies can also be exploited so that the projection and prolongation need only be evaluated locally. We will introduce this technique in the context of hierarchical transformations.

3.6.1. Hierarchical basis. We assume that the linear system (3.1) has been derived from the finite element discretization of an elliptic problem. We also assume that the system of linear spaces (3.24) and the projections I_{k+1}^k have been derived from nested finite element meshes and the accordingly nested finite element spaces. Based on the nesting of spaces, a vector of unknowns x_k on each level k $(2 < k \leq l)$, except the coarsest, can be partitioned as

$$x_k = \begin{bmatrix} x_k^C \\ x_k^F \end{bmatrix}, \tag{3.49}$$

where x_k^C represents the fine grid nodes also present on the next coarser level and x_k^F represents the new nodes on level k. To describe interactions between different levels, we introduce the mapping of indices, $J_{k_1}^{k_2}$,

$$j = J_{k_1}^{k_2}(i), \tag{3.50}$$

where j is the node of a grid k_2 coinciding with node i of grid k_1 when j exists, and is undefined otherwise. We have $x_k^C \in \mathbb{R}^{n_{k-1}}$ and $x_k^F \in \mathbb{R}^{n_k - n_{k-1}}$. Therefore, the interpolation operator I_k^{k-1} has the representation

$$I_{k-1}^k = \begin{bmatrix} I \\ \bar{I}_{k-1}^k \end{bmatrix}.$$

DEFINITION 3.6.1. *The* **hierarchical transformation** $H_k^{k-1} : \mathbb{R}^{n_k} \longrightarrow \mathbb{R}^{n_k}$ *is defined by*

$$x_k = \begin{bmatrix} x_k^C \\ x_k^F \end{bmatrix} \xrightarrow{H_k^{k-1}} \hat{x}^k = \begin{bmatrix} x_k^C \\ \hat{x}_k^F \end{bmatrix} = \begin{bmatrix} I & 0 \\ -\bar{I}_{k-1}^k & I \end{bmatrix} \begin{bmatrix} x_k^C \\ x_k^F \end{bmatrix}. \tag{3.51}$$

The quantities $\hat{x}_k^F$ *are called* **hierarchical displacements.**

Note that the inverse of the hierarchical transformation is given by

$$(H_k^{k-1})^{-1} = \begin{bmatrix} I & 0 \\ \bar{I}_{k-1}^k & I \end{bmatrix}.$$

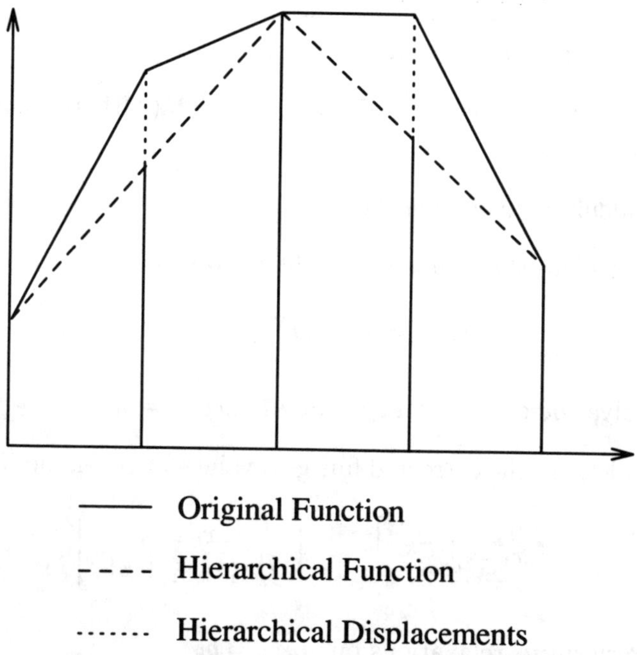

FIG. 3.7. *Hierarchical transformation in one dimension.*

The hierarchical transformation for a one-dimensional example is illustrated in Fig. 3.7.

Now the matrix A_k (interpreted as a bilinear function $A_k : \mathbb{R}^{n_k} \times \mathbb{R}^{n_k} \to \mathbb{R}$) can be partitioned in a compatible form as

$$A_k = \begin{bmatrix} C_k & X_k \\ (X_k)^T & F_k \end{bmatrix},$$

and it can be transformed into

$$\hat{A}_k = \begin{bmatrix} I & (\bar{I}_{k-1}^k)^T \\ 0 & I \end{bmatrix} A_k \begin{bmatrix} I & 0 \\ \bar{I}_{k-1}^k & I \end{bmatrix} = \begin{bmatrix} \hat{C}_k & \hat{X}_k \\ (\hat{X}_k)^T & F_k \end{bmatrix},$$

where

$$\hat{C}_k = C_k + (\bar{I}_{k-1}^k)^T (X_k)^T + X_k \bar{I}_{k-1}^k + (\bar{I}_{k-1}^k)^T F_k \bar{I}_{k-1}^k$$

and

$$\hat{X}_k = X_k + (\bar{I}_{k-1}^k)^T F_k.$$

Note that solving the equation

$$A_k x_k = b_k = I_{k+1}^k b_{k+1} \tag{3.52}$$

is equivalent to solving the transformed equation

$$\hat{A}_k \hat{x}_k = \hat{b}_k = \begin{bmatrix} I & (\bar{I}_{k-1}^k)^T \\ 0 & I \end{bmatrix} b_k. \tag{3.53}$$

1. Perform ν_1 relaxations on $A_k x_k = b_k$.
2. Perform the hierarchical transformation (3.51) to calculate $\hat{x}_k$.
3. Calculate $\bar{\tau}_k$ of equation (3.54).
4. Calculate the coarse grid right-hand side
$$b_{k-1} = b_k^C + (\bar{I}_{k-1}^k)^T b_k^F - \tau_k.$$
5. Solve the coarse grid system $A_{k-1} x_{k-1} = b_{k-1}$.
6. Calculate the corrected fine grid values by reversing (3.51):
$$x_k = \begin{bmatrix} x_k^C \\ x_k^F \end{bmatrix} \longleftarrow \begin{bmatrix} x_{k-1} \\ \bar{I}_{k-1}^k x_{k-1} + \hat{x}_k^F \end{bmatrix}.$$
7. Perform ν_2 relaxations on $A_k x_k = b_k$.

FIG. 3.8. *Basic hierarchical two-level solution algorithm.*

REMARK 3.6.1. *The* coarse grid block $\hat{C}_k$ *in the transformed matrix coincides with the Galerkin coarse grid matrix* A_{k-1}. *If the problem has coefficients that are piecewise constant in the coarse mesh triangulation, then* A_{k-1} *coincides with the finite element stiffness matrix for a coarser problem. This can be seen by interpreting the upper block line of* (3.53) *as the problem of minimizing the energy in the coarse grid finite element space.*

DEFINITION 3.6.2. *The* **τ-terms** *are defined by*

$$\bar{\tau}_k = A_k \begin{bmatrix} 0 \\ \hat{x}_k^F \end{bmatrix} \tag{3.54}$$

and

$$\tau_k = I_k^{k-1} \bar{\tau}_k = I_k^{k-1} A_k x_k - A_{k-1} x_{k-1}. \tag{3.55}$$

REMARK 3.6.2. *The term* τ *defined in equation* (3.55) *is equivalent to the* τ*-term of Brandt* [28], *and* $\bar{\tau}_k$ *is its fine grid representation. The* τ*-term is a correction to the right-hand side of the coarse grid equation such that its solution coincides with the fine grid solution. A multigrid algorithm can be interpreted as an iterative method to determine the* τ*-term.*

The hierarchical transformation leads naturally to a two-level solution technique, and, if applied recursively, to a multilevel iteration. We perform a multilevel algorithm, consisting of smoothing steps alternating with coarse grid corrections. The algorithm is displayed in Fig. 3.8, where step 5 must be replaced by a recursive application of the same algorithm to obtain a V-cycle. A description of this so-called hierarchical transformation multigrid method can be found in Griebel [40] and McCormick and Rüde [59]. A similar approach has been discussed by Mitchell [64], [65].

REMARK 3.6.3. *This algorithm is fully equivalent to a regular (variational) multigrid algorithm and reduces to the special case of Simple-V-Cycle in Fig. 3.5, when just one step of presmoothing and no postsmoothing is applied.*

REMARK 3.6.4. *Though we use hierarchical transformations to implement a multilevel algorithm, our method is different from the* hierarchical basis multigrid *algorithm suggested by Bank, Dupont, and Yserentant* [8], *because we apply the smoother to all nodes of a level (not just the F-nodes).*

3.6.2. Efficient implementation of the multilevel adaptive iteration by hierarchical transformations.

The purpose of this section is to show how all the calculations occurring in the hierarchical transformation multigrid algorithm, as described in the previous subsection, may be restricted to subsets of nodes. It will be demonstrated that, within a cycle and on all levels, the calculations need only be performed on a subset of the points that depends on what was modified on the next coarser or next finer levels.

This is accomplished by using additional attributes for the nodes. The information on which changes require a recalculation must be propagated from fine to coarse and from coarse to fine levels. For the algorithm MLAI of §3.4 we have introduced the sets U_k and V_k to pass the information on where relaxation must be applied. In this algorithm, however, the transfer of numerical quantities between the levels had not been explicitly specified, and, in particular, we did not attempt to describe how the transfer operations are applied only locally. In this section, we will develop this step in detail and show that the transfer of information between levels can also be restricted to appropriate subsets. For this purpose, the system of two sets per level (U_k and V_k) must be extended. We will continue to use the *active set* $\tilde{S}_k$.

The functionality of the sets U and V of Fig. 3.6 is taken over by the sets RP, RT, RR, RC, and RI, described with $\tilde{S}$ as follows.

- $\tilde{S}$, the *active nodes.* This set is used and updated during relaxation. It is initialized with the nodes that are modified by correction when the algorithm is proceeding from coarse to fine levels and by the nodes whose right-hand side is modified when the algorithm is proceeding from fine to coarse levels.

- RP, the set of nodes where the prolongation to the next finer level must be re-evaluated. This set collects all points that have been changed (by either relaxation or coarse grid correction) since the last refinement.

- RT, the set of nodes where the hierarchical transformation (3.51) must be updated. RT contains only F-nodes. A node is added to RT when its own value is changed, or when a neighboring C-point is changed (in relaxation).

- RR, the set of nodes where the τ-term (3.54) must be recalculated. RR is obtained by adding all connections to RT: $RR = RT \cup \text{Conn}(RT)$.

- RC, the set of nodes (on the coarser grid) where the restriction of the right-hand side and τ-term to the coarse grid must be recalculated. RC_{k-1} reflects RR_k on the coarser grid and is obtained by finding the coarse grid equivalents of all points in $RC_{k-1} = I_k^{k-1}(RR_k \cup \text{Conn}_k(RR_k))$. Here Conn_k denotes all *connections* on level k (see Definition 3.1.1 and the discussion in Chapter 4).

- RI, the set of nodes (on the finer grid) where the interpolation must be performed, that is, where the hierarchical transformation must be reversed.

In practice, it turns out that three sets are enough for storing the above information at any time in the algorithm. One set stores the nodes that participate in a transfer of information to coarser levels, another one collects all nodes that have a changed value and must be considered in the next prolongation to the finer grids, and, as before, a third collects the active nodes in the relaxation. However, for presentation, we will continue to use the six sets introduced above. The dependency of the sets is illustrated in Fig. 3.9. The Boolean function is_C_point()[3] tests whether a point of the fine mesh is a C-node, that is, whether it is also present on the coarser mesh.

These sets are used to trace changes and find the subdomains where the corresponding quantities must be recalculated to maintain consistent states. This assumes that at each node i on level k the intermediate quantities occurring in the multigrid algorithm remain stored. These quantities are:

- the current approximate solution $(x_k)_i$,
- the current scaled residual $(\bar{r}_k)_i$,
- the hierarchical transformation for F-nodes on each level $(\hat{x}_k)_i$ (hierarchical displacement),
- the τ-term $(\bar{\tau}_k)_i$,
- the FAS right-hand side $(b_k)_i$.

These values are reused as long as they remain consistent with the present stage of processing.

We now formally define *consistency conditions* in the algorithm. Each state of processing in our algorithm can be described by the current value of the variables. These variables are related by certain equations or inequalities.

[3]We use this function to distinguish different operations for different types of nodes: *replicated* and *interpolated*. If these node types are implemented as separate *object classes* the distinction between them is already implicitly available by *operator overloading* and *virtual functions*; see §4.6. Thus functions like is_C_point will not be needed because their functionality is already contained in the class hierarchy and the correct operation will be used automatically.

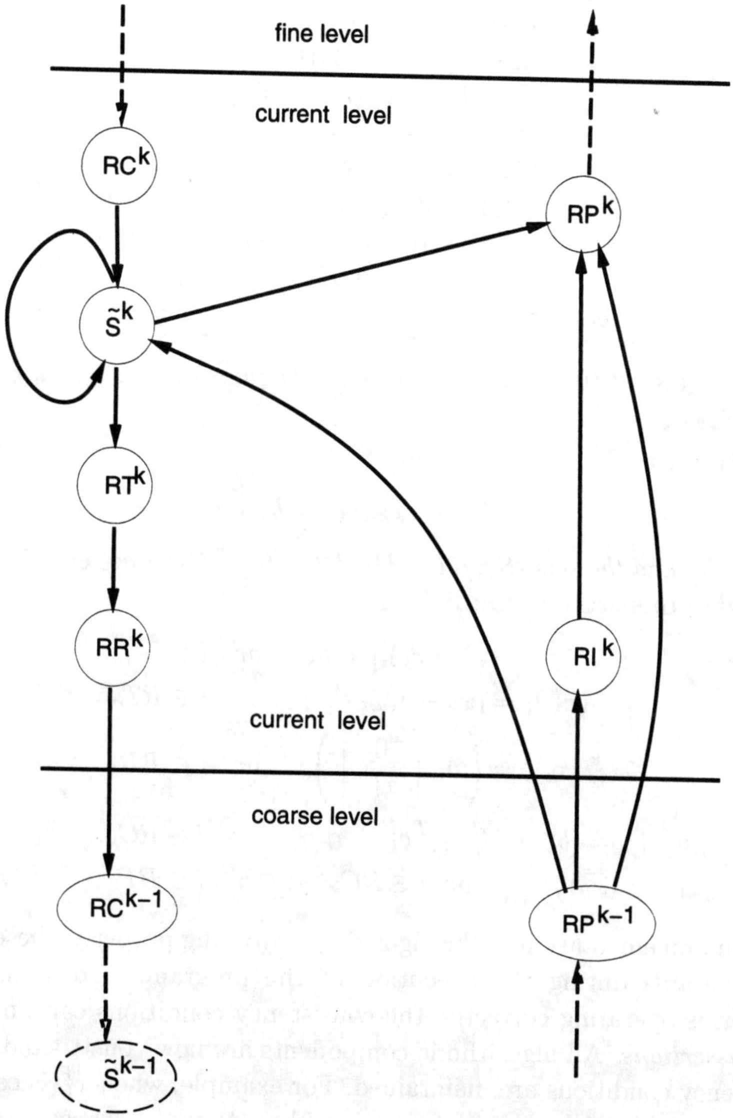

FIG. 3.9. *Interdependency of active sets.*

The goal of the algorithm is to compute a state where these relations are satisfied. When the process has not yet terminated, some of the relations are violated, and the indices of equations, where this is the case, are stored in the sets we have introduced above. Therefore, either the corresponding equation or inequality is satisfied, or the index where it is violated is stored. This leads to the consistency conditions as follows.

```
proc AdaptiveJacobi( ϑ_k, x_k, S̃_k )
    assert((3.56–3.60))
    S = {i ∈ S̃_k| (r̄_k)_i > ϑ_k}
    x_k ← x_k + ω Σ_{i∈S}(r̄_k)_i e_i
    S̃_k ← Conn_k(S)
    RP_k ← RP_k ∪ S
    RT_k ← RT_k ∪ (S ∪ Conn_k(S)) ∩ F_k
    assert((3.56–3.60))
end proc
```

FIG. 3.10. *Augmented adaptive Jacobi relaxation routine.*

DEFINITION 3.6.3. *The vectors*

$$\bar{r}_k, x_k, \hat{x}_k, \bar{\tau}_k, b_k,$$

$1 \le k \le K$, *and the sets* $(\tilde{S}_k, RT_k, RI_k, RR_k, RC_k, RP_k)$ *are* **consistent** *if they satisfy the* **consistency conditions**

$$|(\bar{r}_k)_i| < \vartheta_k \quad or \quad i \in \tilde{S}_k, \tag{3.56}$$

$$(\hat{x}_k^F)_i = (x_k - \bar{I}_{k-1}^k x_k^C)_i \quad or \quad i \in RT_k \quad or \quad i \in RI_k, \tag{3.57}$$

$$(\bar{\tau}_k)_i = \left(A_k \begin{bmatrix} 0 \\ \hat{x}_k^F \end{bmatrix} \right)_i \quad or \quad i \in RR_k, \tag{3.58}$$

$$(b_{k-1})_i = (b_k^C + (\bar{I}_{k-1}^k)^T b_k^F - \tau_k)_i \quad or \quad i \in RC_k, \tag{3.59}$$

$$(x_{k-1})_i = (x_k)_{J_{k-1}^k(i)} \quad or \quad i \in RC_{k-1} \quad or \quad i \in RP_{k-1}. \tag{3.60}$$

If an implementation of the algorithm is working properly, these conditions are *invariants* during the execution of the program. To assure that the program is operating correctly, the consistency conditions can be checked by using *assertions.* All algorithmic components are now constructed so that the consistency conditions are maintained. For example, when $\hat{x}_k^F$ is calculated by performing a hierarchical transformation, the τ-terms may become *inconsistent* for some nodes according to equation (3.58). This must be compensated for by adding to the set RR_k those nodes whose τ-term must be recalculated on the level k.

To prove the correctness of an implementation we now have to check that after each elementary step in the algorithm the status of the algorithm is consistent.

Both the sequential adaptive relaxation scheme of Fig. 3.1 and the simultaneous adaptive relaxation scheme of Fig. 3.4 can be extended to satisfy the consistency conditions. To keep the presentation simple, the following description will only use a single application of adaptive Jacobi as a smoother. We will describe the algorithmic components in detail.

The following subroutines implement the hierarchical transformation multigrid algorithm, each augmented to exploit locality and to maintain consistency.

```
proc HierarchicalTransformation(k)
        assert(3.56–3.60)
        for all i ∈ RT_k
                calculate
                        (x̂_k)_i ⟵ (x_k)_i − [(Ī^k_{k−1}) x^C_k]_i
                RR_k ⟵ RR_k ∪ {i} ∪ Conn_k(i)
        end for
        RT_k ⟵ ∅
        assert(3.56–3.60)
end proc
```

FIG. 3.11. *Calculation of hierarchical transformation.*

Adaptive Jacobi. The adaptive Jacobi relaxation scheme of Fig. 3.10 is an extension of the algorithm in Fig. 3.4. The code in Fig. 3.10 is augmented to manipulate RP and RT. The set F_k denotes the *fine grid nodes* of level k, that is, the *interpolated nodes*; see Chapter 4. The sets RP and RT are used to trigger the correct execution of later restriction and interpolation. Points that are changed must be marked in RP, so that their new value will be transferred to the next finer grid. Depending on whether a changed point is a C-point or not, the node itself, or its neighbors, must be marked for a recalculation of the hierarchical transformation. This in turn determines where coarse grid quantities will change.

Hierarchical transformation. The hierarchical transformation (Fig. 3.11) only needs to be applied to RT, because, according to consistency condition (3.57), the stored hierarchical displacement is still correct at all other nodes. Where the hierarchical displacement is calculated, the residuals will change, so the consistency condition (3.58) demands that these points and their neighbors must be appended to RR.

τ-term. For all points in RR, the τ-term will have changed by way of a modified hierarchical displacement in its neighborhood (see Fig. 3.12). At all other nodes, a previously calculated value may be used. When the changed τ-term is updated, the coarse grid nodes that are affected in a subsequent restriction must be marked in RC.

Restriction. The restriction is performed for all coarse grid nodes in RC where the τ-term has changed (see Fig. 3.13). At all other points, the old values can be reused. A subsequent relaxation is prepared by marking all points with modified right-hand sides in $\tilde{S}$. Depending on whether the point is a C-point or not, the hierarchical displacements at the point or its neighbors must be recalculated, and the point is correspondingly appended to RT.

```
proc CalculateTauterm(k)
        assert((3.56–3.60))
        for all i ∈ RR_k
                calculate (τ̄_k)_i
                if( is_C_point(i) )
                        RC_{k-1} ⟵ RC_{k-1} ∪ {J_k^{k-1}(i)}
                else
                        for all j in Conn_k(i)
                                if( is_C_point( j ) )
                                        RC_{k-1} ⟵ RC_{k-1} ∪ {J_k^{k-1}(j)}
                                end if
                        end for
                end if
        end for
        RR_k ⟵ ∅
        assert((3.56–3.60))
end proc
```

FIG. 3.12. *Recalculation of residuals.*

```
proc Restrict(k)
        assert((3.56–3.60))
        S̃_{k-1} ⟵ S̃_{k-1} ∪ RC_{k-1}
        for all i ∈ RC_{k-1} ( i is a coarse grid node )
                get new approximate solution for i from finer level:
                        (x_{k-1})_i ⟵ (x_k)_{J^k_{k-1}(i)}
                get new right hand for i by restricting
                        the residuals from the finer level:
                        (b_{k-1})_i ⟵ (b_k^C + (Ī^k_{k-1})^T b_k^F − τ_k^{k-1})_i
                if( is_C_point( i ) )
                        RT_{k-1} ⟵ RT_{k-1} ∪ Conn_{k-1}(i)
                else
                        RT_{k-1} ⟵ RT_{k-1} ∪ {i}
                end if
        RC_{k-1} ⟵ ∅
        assert((3.56)–(3.60))
end proc
```

FIG. 3.13. *Restriction of residuals.*

```
proc Correction(k)
    assert((3.56)–(3.60))
    for all i ∈ RP_{k-1} (i is a node on the coarse grid)
        transport solution value to corresponding fine grid node
            (x_k)_{J^k_{k-1}(i)} ⟵ (x_{k-1})_i
        S̃_k ⟵ S̃_k ∪ {J^k_{k-1}(i)}
        RP_k ⟵ RP_k ∪ {J^k_{k-1}(i)}
        RI_k ⟵ RI_k ∪ Conn_k(J^k_{k-1}(i))
    end for
    for all i ∈ RI_k
        reverse the hierarchical transformation
            (x^F_k)_i ⟵ (x̂^F_k)_i + (Ī^k_{k-1} x_{k-1})_i
        S̃_k ⟵ S̃_k ∪ {i}
        RP_k ⟵ RP_k ∪ {i}
    end for
    RI_k ⟵ ∅
    assert((3.56)–(3.60))
end proc
```

FIG. 3.14. *Interpolation of corrections to the fine grid.*

Interpolation. Interpolation (Fig. 3.14) to the fine nodes is performed for all nodes in RP. Within the loop, nodes that are changed on the fine grid are marked for subsequent relaxation sweeps, and are marked for further transfer of the information to still finer levels. Additionally, at the neighbors of these points, the hierarchical transformation must be reversed to obtain new approximate values for the solution. This is stored in set RI. In a second loop, inversion of the hierarchical transformation is evaluated. Values changed in this loop additionally effect the finer grids, and must therefore be marked in RP.

V-Cycle. The subroutines are combined to form a multilevel cycle; see Fig. 3.15. The structure is the same as in a conventional multigrid cycle. The assertions state that the consistency is maintained and that at the start of the algorithm all information must be interpolated to the current level. This V-cycle will usually be repeated until $\tilde{S}_k = \emptyset$ for all k. Note that later V-cycles will usually be executed on progressively smaller sets of nodes.

All these features and statements can be combined to conclude that the complete algorithm in Fig. 3.15 will maintain consistency.

THEOREM 3.6.1. *The multilevel adaptive V-cycle algorithm defined in Fig. 3.15 (and Figs. 3.11–3.14) maintains the consistency conditions.*

Proof. This follows from the construction of the procedures in Figs. 3.10–3.15; see the discussion above.

```
proc V-Cycle( k, S̃1, ϑ1, S̃2, ϑ2, ... )
assert((3.56)–(3.60))
assert( RIk = ∅ and RPk = ∅ for ν = 1, 2, ..., k − 1 )
if ( k = 1 ) then
        solve exactly
else
        AdaptiveJacobi( ϑk, xk, S̃k )
        HierarchicalTransformation(k)
        CalculateTauterm(k)
        Restrict(k)
        V-Cycle( k − 1, S̃1, ϑ1, S̃2, ϑ2, ...)
        Correction(k)
end if
assert((3.56)–(3.60))
end proc
```

FIG. 3.15. *Generalized adaptive V-cycle.*

Theorem 3.6.1 shows that the algorithm, as developed in Figs. 3.10–3.15, maintains a *consistent state* in the sense that the local relations between the variables are satisfied. The algorithm defined by these procedures is a correct implementation of the multilevel adaptive iteration. This makes the error estimates and the work estimates applicable. After each V-cycle, the error can be estimated by examining the residuals. The rate of convergence is determined by Lemma 3.1.1 and the work is determined by Lemma 3.1.3.

3.7. Adaptive iteration on virtual global grids

The adaptive iteration itself does not waste work for unknowns that are already well determined by interpolation from a coarser grid. The adaptive relaxation on the finest level is an error estimator in the sense that it automatically selects the nodes to improve the approximation quality significantly. This is further exploited by the *virtual global grid* refinement technique, which relies on the adaptive iteration as a solver and error estimator. Grids are always refined globally, at least in principle, or *virtually.* If the accuracy at any given point is sufficient, its value can be determined by interpolation from a coarser grid, so no information need be stored there. Generally, memory allocation for a node is deferred until it is really, if ever, accessed and modified by adaptive relaxation. Nodes not explicitly stored are called *ghost nodes*; nodes that are allocated and that have explicitly stored values are called *live nodes.*

Unknowns and nodes are therefore *virtual* entities that are physically created only when required by the algorithm. In this sense, the multilevel adaptive iterative method can be seen as an algorithm for solving the linear systems when the number of levels tends to infinity ($K \to \infty$) and where the solution process is driven only by accuracy requirements.

The residuals at live nodes can be interpreted as contributing to the algebraic error, the residuals at ghost nodes as contributing to the discretization error; see also Remark 3.3.5. If a live node is relaxed, its error contribution is eliminated; if a ghost node is set to be live, its contribution to the discretization error vanishes because the space is enlarged, although its residual will contribute to the algebraic error until the node is relaxed.

The multilevel adaptive relaxation scheme and virtual global grid structure are natural in their unified approach to refinement and iteration. The approximation space is determined on a node-by-node basis, driven by accuracy requirements alone. Whether an already existing live node is relaxed, or whether a ghost node must be set to be live first to enlarge the approximation space, is decided based on the potential gain in the error norm. The FAMe is thus a true PDE solver. The data structure is an infinite sequence of globally refined grids, a typical *recursive, infinite* data structure; see, e.g., [13]. The multilevel adaptive iteration scheme then corresponds to a *lazy evaluation* of the structure, where only those parts of the infinite structure whose presence is necessary to satisfy the accuracy requirements are generated.

The multilevel adaptive algorithms can also be interpreted as a global grid solver with restricted accuracy. The virtual global grid structure represents an infinite system of difference or finite element equations. The selection of live nodes defines which of the unknowns require an explicit solution to obtain a certain overall accuracy. Solving this subsystem also provides values for the ghost unknowns through the interpolation process, and can thus be interpreted as an approximate solver for these equations as well.

3.8. Robustness

Because of the adaptivity, the multilevel adaptive iterative algorithms will be robust for a wide class of problems. The termination criterion for each level guarantees that each subprocess performs to its specification and the termination criterion for the full algorithm guarantees that the final result will be accurate, provided the assumptions of Theorem 3.3.1 or its equivalent are satisfied. These assumptions dictate that we must have a suitable multilevel decomposition of the solution space, leading to a well-conditioned additive Schwarz operator. This is related to an approximate orthogonality relation of the subspaces. It is important to realize that, as for classical multilevel algorithms, this is the core feature that makes the algorithm work. However, in contrast to classical multigrid, the algorithm is still somewhat robust if the orthogonality is violated locally. Local perturbations are compensated for by the adaptive relaxation strategy, which can efficiently remove local errors on all levels.

Furthermore, and equally importantly, diagnostic information is available within the algorithm. If the algorithm runs into problems for which it cannot compensate by additional relaxations, the multilevel cycle will lose its efficiency. Relaxation within a level usually does not introduce large errors on other levels, because the corresponding components of the solution space are approximately orthogonal. In the multilevel adaptive iteration scheme, the information on where errors may have increased is safely propagated throughout the multilevel system.

Thus, if orthogonality of the spaces is severly violated, it will become obvious in the performance of the algorithm. Cycling between levels would not efficiently reduce the size of the active sets, because the removal of errors by relaxation on one level feeds errors on another level. This would suggest that the space decomposition is not optimal, that the estimates of Theorem 3.3.1 are violated (or hold with large c_2/c_1 only), and thus also that the error bound for the solution would be poor, as determined by Corollary 3.3.2. The setup of the multilevel hierarchy (that is, the system of projection and prolongation operators) would thus not be successful and would have to be modified. In contrast to standard multilevel iteration, this can now be easily monitored.

3.9. Parallelization

Up to this point, we have presented a sequential algorithm. Note, however, that the framework in which this algorithm has been derived also allows for parallel implementations. The graph of dependencies in Fig. 3.9 (and the consistency conditions) permit many different solution strategies, including highly parallel ones. The active sets can be interpreted as a queue of orders that can be executed by many processors in parallel. This leads directly to a high degree of *data flow* parallelism.

In practice, we propose implementing parallel algorithms following the usual domain decomposition strategy, using a partitioning of a coarse mesh as the basis of a parallelization. For efficiency reasons, on distributed memory parallel computers, we may partition the orders by subdomains. Each subdomain in turn is associated with one processor so that each processor only executes the computations belonging to its subdomain, using the sequential algorithm developed above. The corresponding data can be stored in the local memory of that processor. Additionally, the dependencies between the subdomains must be traced correctly.

In contrast to conventional parallel multilevel algorithms, the processors need not be synchronized explicitly, but can work fully asynchronously in parallel. In particular, it is not necessary to synchronize between levels, so that different processors may be working simultaneously on different levels of the system. This may cause some loss in efficiency, because data that the processor should have received from the neighboring subdomains may be inconsistent. The dependencies between subdomains may be resolved only after some delay, but the active set concept guarantees that they eventually will be resolved to satisfy overall consistency. This is easily guaranteed if the dependency of the active sets is traced correctly between subdomains. The parallel algorithm terminates when no processor has large residuals on any level. A special parallel multilevel domain decomposition algorithm fitting into this framework has recently been suggested as a *point-block multilevel method*; see Griebel [42].

Our more general parallel concept leaves much flexibility for adapting the algorithm to a specific computer architecture. For a machine with slow communication we may be able to reduce the amount of communication at the cost of increasing the work in each processor; for an architecture with long communication startup delays we may bundle the communication packages into large bulks before sending them. Note that the solution process in each subdomain will continue to perform iterations that improve the overall solution, even if information from neighboring subdomains is not available. Only if this interdomain information is delayed too much may serious performance degradations occur. Eventually, the subdomain process will even stall, because the local solution has been found and can only be improved further when data from neighboring domains becomes available.

3.10. Numerical examples

3.10.1. Adaptive refinement at a re-entrant corner. Our model problem is Laplace's equation (1.1). We illustrate the process of adaptive refinement in combination with adaptive relaxation and cycling. The model problem has a re-entrant corner singularity and is given by

$$\begin{aligned} \Delta u &= 0 && \text{on } \Omega = (-1,1)^2 \setminus (-1,0)^2, \\ u &= r^{2/3}\sin(2/3\phi) && \text{on } \partial\Omega, \end{aligned}$$

where we represent the solution in polar coordinates (r, ϕ).

It is well known that the singularity has a *pollution effect.* Without special corrections (see Rüde [84], [92]) or local refinement, the accuracy deteriorates *globally* to $O(h^{4/3})$ for an interior angle of $3\pi/2$. Furthermore, it can be observed that the convergence rate of a multigrid algorithm deteriorates. This is especially dramatic for the V-cycle. This deterioration can be compensated for by additional local relaxations in the neighborhood of the singularity; see Brandt [30], Mikulinsky [63], and Stevenson [101]. Therefore, we use an algorithm with *self-adaptive local refinement* and *adaptive relaxation.*

Figure 3.16 shows that we do not use virtual global grids yet; instead we use a *conventional finite element refinement strategy,* as outlined in §4.3. We do, however, use an adaptive relaxation scheme to compensate for the deterioration in the multigrid rate of convergence.

The algorithm proceeds from coarser to finer levels in a nested iteration. According to our error indicator, nodes are introduced if their values change by more than a prescribed *refinement tolerance* in a first relaxation sweep. Only if the change is large is the scaled current residual large and is the error bound significantly improved by introducing this node. Nodes with small change are discarded, because they would not improve the error bound efficiently. From level to level of the nested iteration, the tolerance for introducing nodes is reduced by a factor of two.

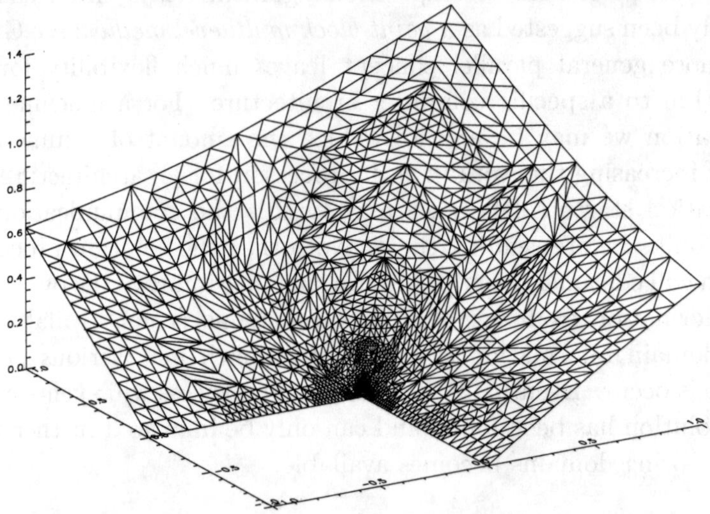

FIG. 3.16. *Locally refined finite element mesh on L-region.*

When the new nodes have been created, a multilevel adaptive relaxation is started. The *relaxation tolerance* for the adaptive relaxation on all levels is determined by the refinement tolerance of the current finest level multiplied by 0.25. This is motivated by the desire to have the algebraic errors somewhat (but not too much) smaller than the differential errors.

A numerical solution on a locally refined finite element mesh is shown in Fig. 3.16. The performance of the algorithm is summarized in Table 3.1. The columns show the number of nodes, the L_2-error after an adaptive multilevel cycle on the given level, the total number of relaxations to obtain this accuracy, and the ratio of the number of nodes to the number of relaxations. The last row displays the $\log_2$ of the progression, averaged over the last five levels. The refinement algorithm is driven by the requirement to reduce the error by a factor of 2 per level. Note that this is reflected by the L_2-error in the second column of Table 3.1. Furthermore, the number of nodes grows with the same factor, as does (approximately) the computational work measured by the number of relaxed points. On the other hand, the number of relaxations per point is still slowly increasing with the number of levels. From the given figures it is impossible to decide whether this relation has an upper bound, or will continue to grow according to some logarithmic dependence on the number of nodes. The cycling structure is visualized in Fig. 3.17.

3.10.2. Virtual global grids and multilevel adaptive solution. Next we study (1.1) with $f(x, y) = 0$ and

$$g(x,y) = \cos(2\pi(x-y))\frac{\sinh(2\pi(x+y+2))}{\sinh(8\pi)}. \tag{3.61}$$

TABLE 3.1
Performance of adaptive multigrid for L-region.

Level	Number nodes	L_2-Error	Number relax.	Relax./Node
1	15	3.654e−02	9	0.60
2	24	1.678e−02	58	2.42
3	41	1.036e−02	151	3.68
4	87	5.941e−03	402	4.62
5	197	2.646e−03	1082	5.49
6	353	1.457e−03	2325	6.59
7	756	7.415e−04	5364	7.10
8	1450	3.990e−04	11161	7.70
9	2830	1.946e−04	22446	7.93
10	5709	1.019e−04	45815	8.03
11	10773	5.243e−05	91996	8.54
rate	0.99	−0.96	1.06	

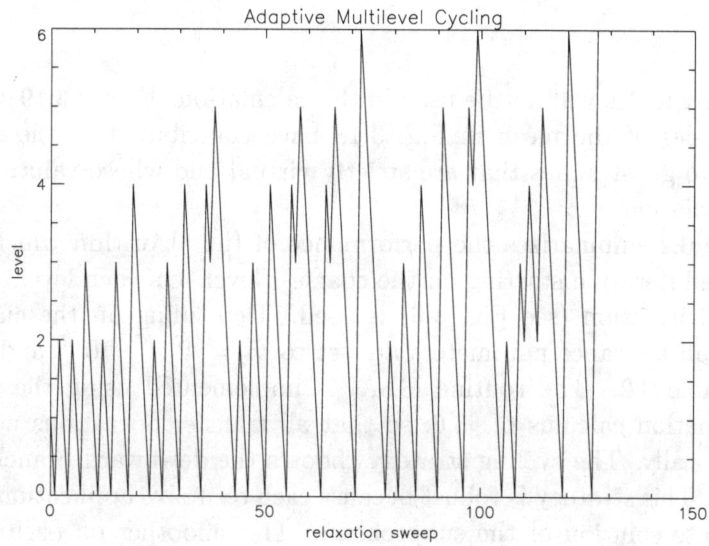

FIG. 3.17. *Cycling strategy in L-region.*

The domain is the square $\Omega = (-1, 1)^2$, and we prescribe Dirichlet boundary conditions $u = g$ on $\partial\Omega = \Gamma$. The analytical solution is known and is given by $u(x, y) = g(x, y)$. The solution to this problem is smooth, $u \in C^\infty(\bar{\Omega})$, but has increased activity in the northeast corner of the domain (see Fig. 3.18).

The coarsest mesh for this example has only one free node at $(0, 0)$ and consists of eight congruent triangles. The mesh of Fig. 3.18 is derived by four subsequent levels of uniform refinement. The finest mesh has 16641 virtual nodes and is obtained by six levels of uniform refinement. Of course, on each

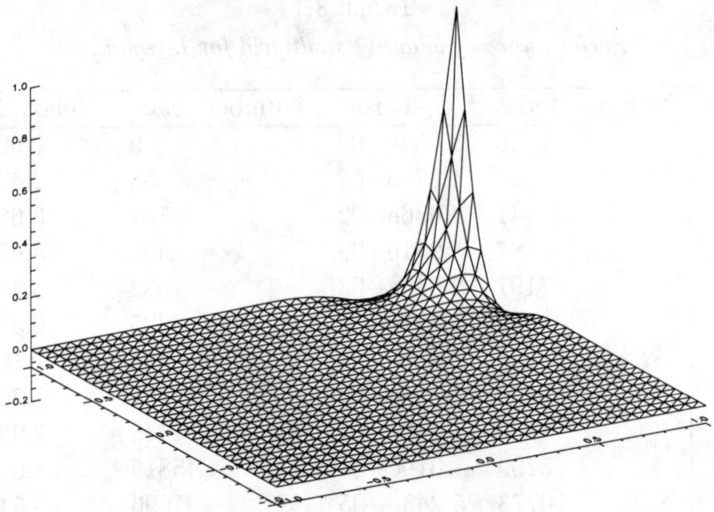

FIG. 3.18. *Solution* $u = g(x, y)$.

level, many nodes will not be used in the calculation. Figure 3.19 shows which of the nodes of the mesh in Fig. 3.18 have contributed to the solution, as opposed to ghost nodes that are strictly virtual and whose values are defined by interpolation.

Table 3.2 summarizes the performance of the algorithm which is applied in a nested iteration starting on the coarsest level. On each level, an adaptive multilevel iteration (see Fig. 3.6) is used. Depending on the current finest level K, all tolerance parameters are set to $\vartheta_k = 4^{7-K} \cdot 10^{-5}$ and designated tol in Table 3.2. The routine MLAI is implemented using the hierarchical transformation calculus of §3.6, so that all transfer operations need only be applied locally. The cycling strategy chooses the downward branch as long as $RT_k \neq \emptyset$. This strategy is robust because each recursive application guarantees an accurate solution of the subproblem. The smoother on each level is the sequential adaptive relaxation scheme of Fig. 3.1.

Each line of Table 3.2 summarizes the status after the termination of the specified level in the nested iteration with the corresponding values of tol. Column "% Active" shows what percentage of the (new) nodes has actually been used. The remaining nodes are ghost nodes. Column "Passes" shows a count of passes through the main loop of the algorithm in Fig. 3.1, and column "Relax" shows how many total relaxation steps have been performed up to this stage. Note that the number of passes is mainly determined by the error estimation process, which has been applied to all nodes. Therefore, the rate of performed relaxations vs. attempted relaxations decreases with the number of levels because it reflects the decreasing number of non-ghost nodes.

Finally, we present the cycling structure generated by the algorithm.

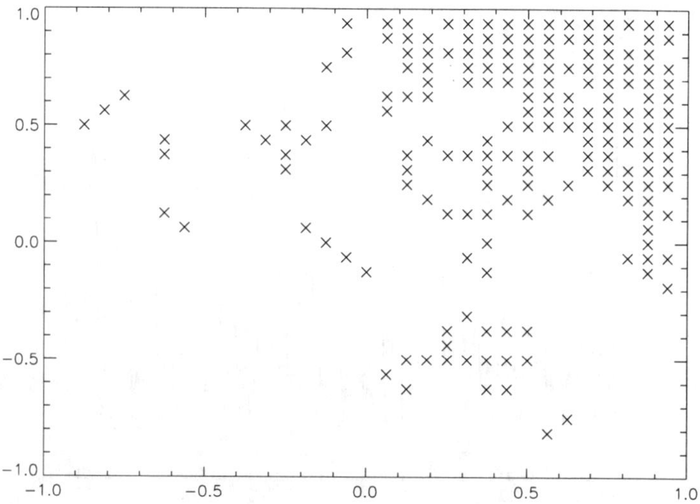

FIG. 3.19. *Active nodes.*

TABLE 3.2
Performance for model problem (3.61).

Level	Nodes	% Active	tol	L_2-Error	Passes	Relax
1	9	100.0	0.04096	0.500000	0	0
2	25	100.0	0.01024	0.013446	123	36
3	81	85.7	0.00256	0.016150	337	63
4	289	65.8	0.00064	0.005652	1423	285
5	1089	47.1	0.00016	0.001504	5741	1143
6	4225	35.5	0.00004	0.000378	21953	3976
7	16641	28.5	0.00001	0.000096	81761	13078

Figure 3.20 shows that the structure is more complicated than in conventional full multigrid (FMG) algorithms. The algorithm forces the residuals to become small on each level before it returns to finer ones. It should be noted, however, that most of the work is accomplished by the first subcycle; subsequent coarse grid processing usually only provides local corrections. Thanks to the locality of all multigrid calculations, later cycles are computationally inexpensive.

3.11. Perspectives

The family of algorithms discussed in this chapter presents only a framework, which can be modified and extended in many ways. The basic idea is not restricted to the assumptions made and can be extended to nonsymmetric and even nonlinear problems. In the following we will discuss a few ideas of how the active set concept can be further developed.

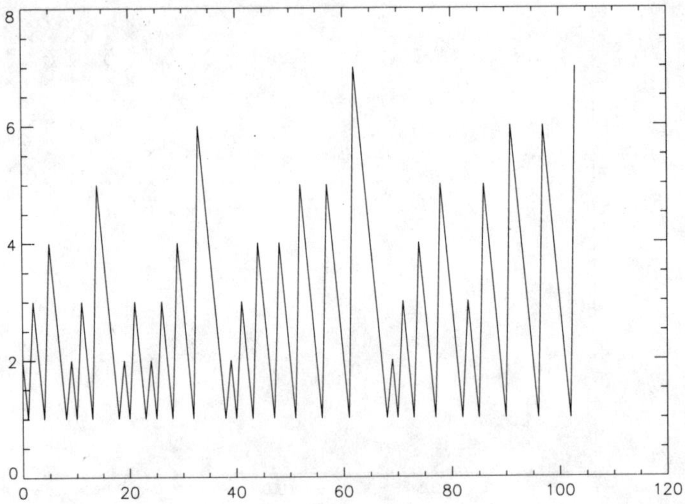

FIG. 3.20. *Cycling structure for model problem* (3.61).

Incremental residual update. With each elementary relaxation, the residual at the *neighboring unknowns* changes. Assume that we store the scaled residuals for each unknown. If equation j is relaxed, then the values and current residuals can be updated according to

$$\begin{aligned} \xi_j &\leftarrow \xi_j + \theta_j, \\ \theta_i &\leftarrow \theta_i + \frac{a_{i,j}}{a_{i,i}} \theta_j, \text{ for all } i \in \text{Conn}(j), \\ \theta_j &\leftarrow 0. \end{aligned}$$

Thus, the residuals are always directly available and can be used to construct the active sets.

Finer active set classification. The present strategy classifies the equations according to whether the residuals are larger or smaller than a single tolerance parameter. It may be useful to make finer distinctions and use several sets, collecting the equations with residuals in a certain range. This would make algorithms that are even closer in spirit to the Gauß–Southwell method feasible. The final tolerance would not be fixed at the beginning, but would slowly be decreased from a starting value to final accuracy. This could be efficiently combined with the idea of incremental computation of the residuals.

Downstream relaxation. Though the present theory from Chapter 2 is limited to symmetric problems, the adaptive relaxation may have specific advantages for other problems, such as convection-dominated equations. For the convection terms, relaxation *in the downstream* direction is necessary for good performance. Adaptive relaxation automatically adapts to such a behavior if the connections are appropriately modeled and used according to their strength. Appropriate extensions to the adaptive relaxation routines are obvious, although at this point we have neither theoretical nor practical results for this type of algorithm.

Transfer of information between levels. When proceeding from coarse to fine levels, the active set strategy is typically much too conservative in constructing the extended active sets, because many fine level nodes depend on few coarse level nodes. It would be desirable to find a more concise strategy. We may try to systematically exploit the fact that the numerical product

$$F = A_k \prod_{\nu=k'}^{k} I_{\nu}^{\nu+1}$$

often has significantly fewer nonzero elements than the corresponding Boolean matrix product that would be used implicitly in our algorithm to determine the active sets of other levels.

Another idea is to exploit the fact that the size of many matrix elements in the transfer operators is usually small, so that coarse level changes usually do not introduce large changes in the finer grid residuals. This could be combined with the idea of multiple active sets.

Grouping unknowns. The strategy of classifying the nodes (or equations) individually may be too expensive. It is possible to group the unknows and change their active/inactive status only collectively. This may be natural when several unknowns are associated with each other, for example, when the matrix originates from a system of PDEs where several variables are associated with a single geometric location. But grouping may also be introduced artificially, e.g., based on domain decomposition. Each such group of unknowns would be either active or inactive and would have its neighbor relations connecting it to other groups.

3.12. Historical remark

At the core of our algorithms are modified relaxations. In contrast to standard relaxation algorithms, like Gauß–Seidel or Jacobi iteration, we exploit the knowledge about residuals to find an ordering of the unknows that leads to faster convergence. Such ideas have been popular since before the advent of modern electronic computers, when relaxation was still performed by humans. We have already cited the papers by Southwell [96], [97], [99], who is also the originator of the term *relaxation*. Actually, however, these ideas are much older. In one of the first references to iterative linear system solvers, a letter of Gauß to Gerling, dated December 26, 1823, Gauß demonstrates the successive relaxation for a 4×4 system using the Southwell ordering. Also, Seidel [94] used the Southwell strategy instead of the algorithm that is now carrying his name. These early references even contain elements that can be regarded as precursors to the active set strategy. From an historical perspective, it may seem that iterative methods, as they have been developed in the past forty years, rely too much on pure computing power. Ideas for improving iterative methods by watching the iteration progress may have been neglected only because it is difficult to code them efficiently. For a summary of the history of iterative methods, see also Ostrowski [67].

Chapter 4

Data Structures for Multilevel Adaptive Methods

4.1. Introduction

4.1.1. Overview. This chapter contains a systematic discussion of data structures for multilevel adaptive techniques.

The *topological* properties of a mesh are discussed in §4.2. The topological structure is successively augmented by *geometric* and *algebraic* information, to provide the basis for the design of practical implementations.

In §4.3 we study several special cases of refined structures. In this section the perspective is *static*, that is, we examine the suitability of mesh structures without discussing the process by which they are generated.

The *dynamic* viewpoint is elaborated in §4.4, where we discuss how the mesh is generated as part of the solution process. The special requirements of the algorithm lead to a hierarchical system of meshes that is studied in §4.5.

Techniques for an *object-oriented* implementation of such structures are finally addressed in §4.6, where we discuss programming techniques suitable for the efficient, high level implementation of numerical algorithms. The programming language C++ provides a basis for implementing adaptive algorithms with advanced software techniques. The material in this chapter is based on the report [86]. For additional information, see Rüde [89].

4.1.2. Relations. Relations play an important role in the specification and construction of software. The mathematical notion of relations is generally well suited for describing data structures and thus we will also use it to discuss adaptive finite element meshes.

A relation R of k sets $M_1, \ldots, M_k$ is a subset of the Cartesian product

$$R \subset M_1 \times \cdots \times M_k.$$

We will primarily use *binary* relations $R \subset M_1 \times M_2$. $(m_1, m_2) \in R$ can then be read as "m_1 *is-related-to* m_2."

Generally, relations have a fixed number of arguments. In our context this is not always suitable because, e.g., an element can be a triangle or a quadrilateral and thus have either three or four edges. Similarly, a node can be the endpoint of any number of edges.

To handle this, we will use the *power set* $\mathfrak{P}(M)$, that is, the set of all subsets of M, and use relations of the form $R \subset M_1 \times \mathfrak{P}(M_2)$. If these relations are unique in the first argument, they can be interpreted as functions:

$$M_1 \xrightarrow{R} \mathfrak{P}(M_2).$$

To this end, the relation may have to be supplemented by adding tuples $(m_1, \emptyset)$, $m_1 \in M_1$, such that the function is defined for all elements of M_1. The augmented relation ($\subset M_1 \times \mathfrak{P}(M_2)$) can be represented by a list of pointers for each element in M_1.

4.1.3. Specification of software. The advantage of an *abstract approach* is that the features of the finite element structure can be discussed and analyzed without referring to a concrete representation. *Elements, edges*, and *nodes* are abstract entities defined only by certain operations and relations between them. We consciously separate the *functionality* required from the *data representation* that implements this functionality.

This software engineering approach is also related to *object-oriented programming*, where a *class* of *objects* is defined by the operations it can perform and not by its internal representation. In object-oriented programming the operations that a *class* can perform and the internal representation of the class are designed separately. From outside, the representation is irrelevant; only the operations that can be performed are of interest. This concept of *data hiding* improves the modularity, flexibility, and extensibility of the software.

Our discussion may also be interpreted as the informal use of *abstract data types*. The informal description could be extended to the definition of a *rigorous abstract data type* that includes a consistent system of *laws* that specifies the data structure axiomatically (see, e.g., Bauer and Wössner [13]). The design of *rigorous* abstract data types is beyond the scope of our presentation and we refer to the theses by Adler [1] and Stark [100], where rigorous approaches are outlined.

Though the use of abstract data types as a *rigorous* software development method for finite element problems is beyond the scope of the present work, we will show that the informal use of these techniques is a useful design method.

4.1.4. Efficiency and modularity. The finite element algorithms determine which operations must be defined for the data type. The optimal internal representation in turn depends on the operations that must be performed with it. In the language of relations one goal may, e.g., be to find a *minimal representation* with as few relations as possible. Here the other relations must be constructed whenever needed.

Though the functionality will be described independently of the internal representation, the design of the internal representation clearly effects the efficiency. Storage requirements must be balanced with run-time performance, and for an optimal design we must carefully analyze the behavior of the operations implemented on that data structure. Of course, different implementations of the same abstract type influence the quality of the software. In §§4.5.2 and 4.5.3 we will, e.g., compare specific advantages of an element-based representation versus a node-based representation of finite element meshes. Both representations can be extended to provide the same abstract operations, but they will differ in how efficiently certain operations are performed.

But the problem is even more complex. The final goal is not to implement finite element meshes, but to solve certain PDEs. For a specific problem, the full flexibility of general finite elements may not be needed. Implementing the full scope of operations and structures necessarily introduces a significant overhead. This is especially true if the program is to be implemented for a high performance computer of vector or parallel type. The performance of these architectures can only be exploited if the data structures have a certain uniformity.

On uniform rectangular meshes certain basic operations are inherently more efficient than on fully adaptive, unstructured grids that create more organizational overhead. Of course, it depends on the problem at hand whether the increased flexibility of unstructured meshes leads to a general solver that is more efficient than an equivalent solution on a more rigid data structure, where the *computational efficiency* may be better, but much work may be wasted because the structure cannot be suitably adapted to the problem features.

For some problems, a uniform mesh may be appropriate for the problem under study, and then a code based on unstructured meshes will not be competitive. This motivates the study of mesh structures with restricted flexibility in §4.3.

Resolving these conflicting goals is a great challenge for the software architecture. Our goal is to develop a concept that includes the uniform, easily vectorizable structures up to the fully unstructured adaptive meshes, yielding optimal computational efficiency in one case, and maximal flexibility in the other. We will show that this can be accomplished by careful modularization using concepts beyond the simple procedural software decomposition based on classical languages. In §4.6 we will show how advanced software engineering techniques can be used to achieve some of these goals.

4.2. Finite element meshes

4.2.1. Classification. The structure of a finite element problem has four different aspects: *topology, geometry, physics*, and *algebra.*

From a *topological* viewpoint a mesh only consists of *elements*, *edges*, and *vertices* (nodes). These are abstract entities that must satisfy certain neighbor relations to guarantee that the mesh is logically consistent.

The *geometric* aspect adds *metric* information to the topology, including the location of vertices, the length of edges, the area of elements, angles, distances, etc. The geometric component is essential for the adaptation process and for obtaining a highly accurate solution.

Another set of information is of a *physical-mathematical* nature. Physical components of the problem description include data associated with the problem to be solved, like material coefficients, and force or source terms.

Using the finite element process we finally derive the *algebraic* aspect as a discrete representation of the physical problem. In this step we introduce *nodal variables* forming a system of unknowns to be solved. The physical equation is transformed into discrete equations that are described by entities like the *stiffness* and *mass matrix*.

This concept of reducing the finite element data structure to a basic topological structure that is successively enriched with geometry and algebra will later be reflected in our object-oriented software design. The topology corresponds to the basic data structure that is extended by geometric and algebraic information through the *inheritance* mechanism discussed in §4.6. The physical components could be added analogously. However, this will not be discussed because physical information is not directly relevant for the solution process that is the main focus of this monograph.

4.2.2. Triangulations. Equation (1.11) can be discretized by finite elements. We assume the polygon Ω is partitioned by a triangulation $(\mathcal{T}, \mathcal{N}, \mathcal{E})$ consisting of t *triangles* and *quadrilaterals*,

$$\mathcal{T} = \{T_1, \ldots, T_t\}.$$

The vertices of all triangles and quadrilaterals together constitute the set $\mathcal{N}$ containing n *nodes* (or vertices) of the triangulation. Finally, there is the set $\mathcal{E}$ of *edges*, consisting of e lines connecting the vertices.

For an example of a finite element function on a triangulated domain, see Fig. 3.16. The formal definition for this situation is given by

$$\begin{aligned}
\mathcal{N} &= \{N_1, \ldots, N_n\} \\
&= \bigcup_{T(N_1N_2N_3)\in\mathcal{T}} \{N_1, N_2, N_3\} \cup \bigcup_{Q(N_1N_2N_3N_4)\in\mathcal{T}} \{N_1, N_2, N_3, N_4\}, \\
\mathcal{E} &= \{E_1, \ldots, E_e\} \\
&= \bigcup_{T(N_1N_2N_3)\in\mathcal{T}} \{\overline{N_1N_2}, \overline{N_2N_3}, \overline{N_1N_3}\} \cup \\
&\quad \bigcup_{Q(N_1N_2N_3N_4)\in\mathcal{T}} \{\overline{N_1N_2}, \overline{N_2N_3}, \overline{N_3N_4}, \overline{N_4N_1}\}.
\end{aligned}$$

The Ciarlet conditions. To be suitable for the finite element process, a triangulation $(\mathcal{T}, \mathcal{N}, \mathcal{E})$ must satisfy the following requirements (see Ciarlet [33]):

T1. The closure of the region Ω is the union of the elements in $\mathcal{T}$,

$$\bar{\Omega} = \bigcup_{T\in\mathcal{T}} T.$$

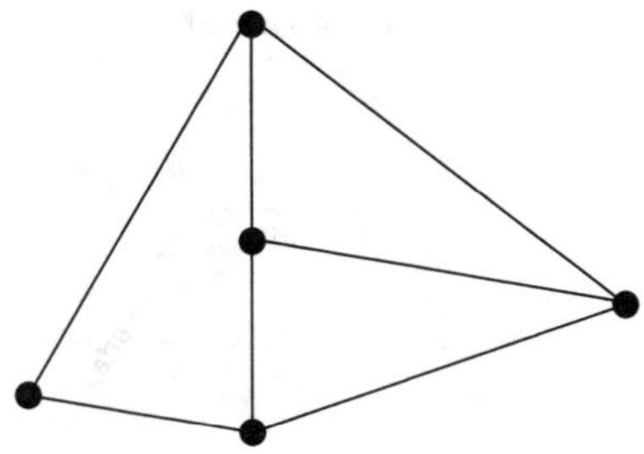

FIG. 4.1. *Nonconforming triangulation.*

T2. Each element $T \in \mathcal{T}$ is closed ($T = \bar{T}$) and its interior is nonempty ($\mathring{T} \neq \emptyset$).

T3. Elements do not overlap, that is, for each distinct pair of elements, $T_1, T_2 \in \mathcal{T}$: $\mathring{T}_1 \cap \mathring{T}_2 = \emptyset$.

Conforming triangulations. We call a triangulation *conforming* if, in addition to T1–T3, condition T4 holds:

T4. Any edge of any element T_1 is either a subset of the boundary $\partial\Omega$ or an edge of another element $T_2 \in \mathcal{T}$.

To be conforming, nodes may only be located at the endpoints of edges.[1] In conforming triangulations two adjacent triangles intersect in a common edge, and any two distinct triangles are either disjoint or they intersect in one common node or one common edge. For an example of a triangulation with a *nonconforming node*, see Fig. 4.1.

In the set of vertices $\mathcal{N}$ we also distinguish between *constrained* vertices $\mathcal{N}_D$ lying on a Dirichlet boundary, and the set of *free* vertices $\mathcal{N}_I$ in the interior of the domain or on a part of the boundary not constrained by a Dirichlet condition.

With the system of sets $(\mathcal{T}, \mathcal{N}, \mathcal{E})$ we can describe the *topological* structure of a mesh if additional conditions are satisfied.

4.2.3. Topological structure of a finite element mesh.

Basic relations. The topology defines neighbor relations between any of the three basic sets $(\mathcal{T}, \mathcal{N}, \mathcal{E})$. Table 4.1 gives an overview of the basic relations; see also Fig. 4.2. Two distinct nodes (vertices) are called *adjacent* (*neighbors*), if they are endpoints of the same edge. Similarly, we will call two triangles T_1

[1]Here we are only discussing topological and geometric aspects. *Algebraic* nodes for higher order basis functions are of course often located on the edges or in the elements.

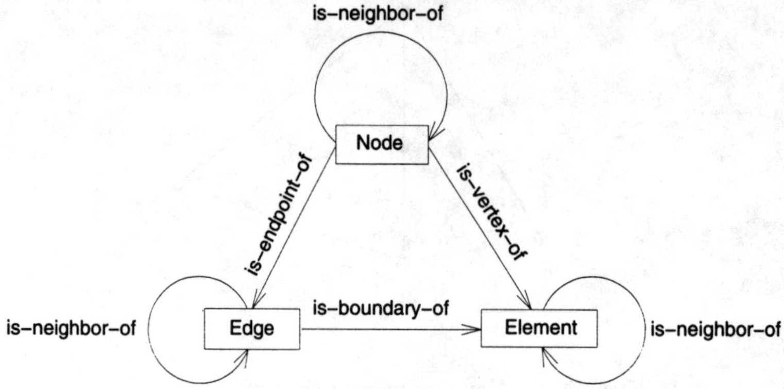

FIG. 4.2. *Relations between basic entities.*

TABLE 4.1
Basic topological relations in a finite element mesh.

Name	Subset of	Description
is-neighbor-of	$\mathcal{N} \times \mathcal{N}$	nodes connected by an edge
is-endpoint-of	$\mathcal{N} \times \mathcal{E}$	endpoints of edges
is-vertex-of	$\mathcal{N} \times \mathcal{T}$	vertices of elements
is-neighbor-of	$\mathcal{E} \times \mathcal{E}$	edges sharing a common node
is-boundary-of	$\mathcal{E} \times \mathcal{T}$	edges that form an element
is-neighbor-of	$\mathcal{T} \times \mathcal{T}$	elements that share a common edge

and T_2 *adjacent* or *neighbors* if an edge of T_1 is contained in T_2 or an edge of T_2 is contained in T_1.

The relations in Table 4.1 describe the topology of the mesh. They are redundant, so that not all must be stored explicitly in representation of the data type. Most of them can be computed from others. This will be discussed further below. Additionally, consistency conditions must be satisfied.

Further relations. From the basic *is-neighbor-of* relation on nodes we define the "*has-neighbors*" relation as a subset of $\mathcal{N} \times \mathfrak{P}(\mathcal{N})$. Similarly, the endpoints of an edge in $\mathcal{E}$ are two nodes in $\mathcal{N}$. This can be expressed as a relation *"has-endpoints"* $\subset \mathcal{E} \times \mathcal{N} \times \mathcal{N}$. Three or four nodes constitute the corners of an element, so that we can introduce the relation *"has-vertices"* $\subset \mathcal{T} \times \mathfrak{P}(\mathcal{N})$. These three relations are unique in the first argument, so that they can be used to define functions.

We will frequently implement these relations as set-valued functions. For computing the set of neighbors of a node we introduce

$$\text{Neigh}(N) = \{N' \in \mathcal{N} | \; \exists (N, N') \in \textit{is-neighbor-of}\}. \tag{4.1}$$

Note also that the *is-neighbor-of* relation can be interpreted as a representation of the *edges*, more precisely,

$$(N, N') \in \textit{is-neighbor-of} \Leftrightarrow \exists E = \overline{NN'} \in \mathcal{E}$$

or

$$(N, N') \in \textit{is-neighbor-of} \Leftrightarrow \exists E \in \mathcal{E} : (N, E), (N', E) \in \textit{is-endpoint-of}.$$

This interpretation is particularly useful if the set of nodes $\mathcal{N}$ is the primary data type, as discussed in §4.5.3.

Similarly, the set of finite elements $\mathcal{T}$ can be interpreted as two relations with three and four arguments, respectively, that represent the elements. This relation is equivalent to the *has-vertices* relation.

If the mesh consists of triangular elements only, the set of elements $\mathcal{T}$ can be interpreted as a ternary relation on the set of nodes $\mathcal{T} \subset \mathcal{N} \times \mathcal{N} \times \mathcal{N}$. Here the set of elements $\mathcal{T}$ is defined by the 3-cycles in the graph of the *is-neighbor-of* relation on the nodes. This can be used to avoid the explicit storage of the $\mathcal{T}$. Conversely, we can construct the edges $\mathcal{E}$ (interpreted as an *is-neighbor-of* relation) as the union of the three different projections of $\mathcal{T}$ to binary relations.

Therefore, if only triangular elements are used, only one of $\mathcal{E}$ and $\mathcal{T}$ (interpreted as relations) must be stored explicitly; the other one can be constructed. However, if we include other (quadrilateral) elements, and perhaps generally, for performance reasons, both relations should be stored.

An *interior* edge is the boundary between two finite elements, so that we have a relation *"is-boundary-between"* $\subset \mathcal{E} \times \mathcal{T} \times \mathcal{T}$. The edges on the boundary of the domain are exceptions, because they are the boundary of only one element. These edges may be included in the above concept if we interpret the exterior of the domain (and holes in the domain, if present) as a special element.

These relations define the essence of the topological structure of a finite element mesh $(\mathcal{T}, \mathcal{N}, \mathcal{E})$. As remarked above, a mesh with triangles only is already uniquely determined, if only the set $\mathcal{N}$ and the *is-neighbor-of* relation are stored for the nodes. From there we can build all the other relations, like *is-boundary-between.*

Euler's formula. For conforming triangulations of a simply connected region (without holes) the Euler theorem for polyhedra can be applied and yields

$$n - e + t = 1, \tag{4.2}$$

where n, e, and t are the cardinalities of the sets $\mathcal{N}$, $\mathcal{E}$, and $\mathcal{T}$, respectively. The formula can be generalized to include more complicated situations, like domains with holes; see, e.g., Stark [100].

The basic importance of the Euler formula motivates the construction of finite element meshes by the so-called *Euler operators.* The Euler operators manipulate a mesh in such a way that consistency with respect to equation

(4.2) remains satisfied. This technique was originally developed for geometric modeling and computer-aided design. The suitability of Euler operators for the finite element mesh construction is studied in Adler [1] and Stark [100]. The Euler operators are also suitable for a rigorous specification of the topological structure and its definition as an *abstract data type.*

4.2.4. Geometric structure. If the topological structure of a finite element mesh is determined, the geometric information can be stored by augmenting the set of nodes with *attributes* containing their geometric location. All other geometric information needed, like distances, lengths, and areas, can be simply computed from there by basic analytic geometry. This can be implemented in the data structures by extending the node data type by suitable components.

Leinen argues in [49] that for reasons of numerical stability the length of the edges should be stored explicitly. The continued halving of a real number is less sensitive to roundoff errors than the calculation of the distance of nodes that have almost the same coordinates. Similarly, it may be useful to store the direction of an edge explicitly—if necessary, in the algorithms.

In the finite element context, a mesh must satisfy certain geometric conditions, as they are partly contained in the Ciarlet conditions T1–T4. Additional constraints are introduced below.

Regular families of triangulations. An element $T \in \mathcal{T}$ is characterized by

$$\sigma(T) := \frac{\operatorname{diam}(T)}{\rho(T)},$$

where

$$\rho(T) = \sup\{\operatorname{diam}(S) |\ S \text{ ball contained in } T\},$$

and where the diameter of a set $S \subset \mathbb{R}^n$ is defined by

$$\operatorname{diam}(S) = \sup\{\operatorname{dist}(a,b) |\ a, b \in S\}.$$

For a triangulation $\mathcal{T}$ we define

$$\sigma(\mathcal{T}) = \max\{\sigma(T) |\ T \in \mathcal{T}\},$$

and the *characteristic mesh size*

$$h(\mathcal{T}) = \max\{\operatorname{diam}(T) |\ T \in \mathcal{T}\}.$$

Usually, the individual members of a family of triangulations are characterized by their values, $h = h(\mathcal{T})$, and consequently we denote them by introducing the superscript h: $\mathcal{T}^h$, $\mathcal{N}^h$, and $\mathcal{E}^h$.

A family of triangulations $(\mathcal{T}^h, \mathcal{N}^h, \mathcal{E}^h)$ is called *regular* if the following two conditions are satisfied (see Ciarlet [33]):

T5. $h(\mathcal{T}^h) \to 0$,

T6. $\sigma(\mathcal{T}^h) \leq \sigma$ for all h with σ independent of h.

Condition T6 is satisfied if the minimal angle of all triangles is bounded from below independently of h. For a discussion of this *minimal angle condition* and its replacement by the weaker *maximal angle condition*, see Babuška and Aziz [3]. With angles larger than 90 degrees, we lose the M-matrix feature of the stiffness matrix.

4.2.5. Algebraic structure. When the topology and the geometry of a mesh are determined, the algebraic structure of the finite element problem is derived from the physical and mathematical problem description. This is discussed in this section.

Finite element spaces associated with a triangulation. We will concentrate on the simplest type of element. The finite element space $V^h = V^h(\mathcal{T}^h)$ for triangular elements is defined by

$$\begin{aligned} &V^h \subset H^1(\Omega), \\ &u \in V^h \Rightarrow u \qquad \text{linear in } T_i^h \text{ for } T_i^h \in \mathcal{T}^h. \end{aligned}$$

If parallelogram elements are included, we must allow *bilinear* functions for them. For general quadrilaterals we must even introduce isoparametric elements (see Schwarz [93]). In the space V^h we define *nodal basis functions* B_i^h by

$$(4.3) \qquad \begin{cases} B_i^h \in V^h, & & \\ B_i^h(x,y) & = & 0 \quad \text{for } (x,y) \in \mathcal{N}^h, \quad (x,y) \neq N_i^h, \\ B_i^h(x,y) & = & 1 \quad \text{for } (x,y) = N_i^h. \end{cases}$$

Clearly, the support of a nodal basis function is given by

$$(4.4) \qquad \text{supp } B_i^h = \bigcup_{N_i \in T_i^h} T_i^h.$$

For homogeneous Dirichlet boundary conditions we have $g = 0$ and we must study the space $H_0^1(\Omega)$ and its subspaces V_0^h. The space V_0^h is spanned by the basis functions

$$(4.5) \qquad V_0^h = \text{span}\{B_i^h | N_i^h \in \mathcal{N}_I^h\}.$$

For the solution of (1.10) with $g \neq 0$ we introduce the finite-dimensional *affine* space V_g^h

$$V_g^h = V_0^h + \sum_{N_i^h \in \mathcal{N}_D^h} g(N_i^h) B_i^h,$$

where the fixed linear combination of functions associated with constrained nodes enforces the Dirichlet boundary conditions at the nodal values on the boundary. Usually, $V_g^h \not\subset H_g^1(\Omega)$. The functions in V_g^h satisfy the boundary conditions only in the nodal values.

Discretized problem. Each function in V_g^h is consequently characterized by a vector of $n_I^h = |\mathcal{N}_I|$ real values corresponding to its representation in the above *nodal* basis. A numerical approximation to equation (1.10) or (1.11) is obtained by replacing the continuous space $H(\Omega)$ by its discrete approximation and by solving the finite-dimensional minimization problem

$$a_\Omega(u^h, v^h) - 2(f, v^h)_\Omega = 0 \tag{4.6}$$

for all $v_h \in V_0^h$. This can be generalized by using different test and trial spaces. Representing u^h with respect to the nodal basis with coefficient vector

$$\vec{u} = (u_m)_{m=1,\ldots,n_I^h} \in \mathbb{R}^{n_I^h}$$

leads to a linear system

$$A^h \vec{u}^h = \vec{f}^h, \tag{4.7}$$

where A^h is the so-called *stiffness matrix* with the components

$$a_{i,j} = a_\Omega(B_i^h, B_j^h) = \iint_{\mathrm{supp}B_i \cap \mathrm{supp}B_j} (\nabla B_i^h)^T D(x,y) \nabla B_j^h \, dxdy \tag{4.8}$$

for model problem (1.2).

Edge-oriented representation. Note that for triangular elements, $a_{i,j} = 0$, if $\overline{N_i N_j} \notin \mathcal{E}$. Therefore, if A^h is symmetric, its components can be represented as attributes of the edges. This is an appropriate data structure if $\mathcal{E}$ is stored as a primary data type.

For nonsymmetric operators this must be generalized by providing two attributes per edge of the triangulation or by introducing a directed graph. Here the edges will be called *semi*-edges. This is appropriate if the semi-edges are stored as components of a node data structure. Each real edge $\overline{N_i N_j}$ is then represented by two attributed semi-edges

$$(\overline{N_i N_j}, a_{i,j}) \quad \text{and} \quad (\overline{N_j N_i}, a_{j,i}). \tag{4.9}$$

$a_{i,j}$ and $a_{j,i}$ can be associated with an edge in $\mathcal{E}$. This is a special case depending on the use of triangular elements. If bilinear, quadrilateral elements are used, nodal values at diagonally opposite nodes of the quadrilateral are coupled by nonzero entries in the element matrix. To keep the data structure extensible for more general elements, it is therefore advisable to treat the topological relations (edges) and the algebraic ones (connections) separately.

We introduce the binary relation *"is-connected"* $\subset \mathcal{N} \times \mathcal{N}$. The set of **connections** is

$$\mathcal{C} = \{(N_i, N_j, a_{i,j}) |\; a_{i,j} \neq 0\}. \tag{4.10}$$

For triangular elements this is the same as the semi-edge relation introduced above and need not be stored separately. For general elements we must distinguish between *geometric* and *algebraic* neighbors. We further introduce the set of *connections of a node* N_i (see also Definition 3.1.1):

$$\operatorname{Conn}(N_i) = \{N_j |\ i \neq j,\ a_{i,j} \neq 0\}.$$

The *connections* are the basis for implementing efficient sparse matrix multiplication and relaxation algorithms.

Matrix-oriented representation. For uniform or quasi-uniform meshes on a (logically) rectangular domain, a lexical ordering of the nodes produces a banded matrix that may be stored efficiently in terms of the bands. More general structures may be stored as *profiled* matrices.

This may be appropriate when all the storage within the matrix profile will be needed anyway, because a direct solver generates fill-in at these positions. Sparse matrix techniques appropriate for iterative methods (which are our main interest) often use lists to represent the sparsity structure and are therefore similar to the graph-oriented techniques using connections. A list containing the nonzero entries for row i of A^h is a representation for the set $\operatorname{Conn}(N_i)$ and thus such a structure is a representation of the connection relation.

Element-oriented representation. Another technique to represent A^h is derived directly from the usual process of its computation. For each $T = \triangle(N_i, N_j, N_k) \in \mathcal{T}$, we can define the *element stiffness matrix* $A_{\triangle(N_i,N_j,N_k)}$ as the 3×3 matrix

$$A_T = \begin{bmatrix} a_T(B_i^h, B_i^h) & a_T(B_i^h, B_j^h) & a_T(B_i^h, B_k^h) \\ a_T(B_j^h, B_i^h) & a_T(B_j^h, B_j^h) & a_T(B_j^h, B_k^h) \\ a_T(B_k^h, B_i^h) & a_T(B_k^h, B_j^h) & a_T(B_k^h, B_k^h) \end{bmatrix},$$

where the integrals in (4.8) are calculated for the triangle T only. This must be generalized to 4×4 matrices for bilinear elements, and to still larger matrices for more complicated elements, to represent systems of PDEs or higher order elements.

The stiffness matrix A^h is *assembled* from the element stiffness matrices in the form

$$A^h = \sum_{T \in \mathcal{T}^h} \Pi_T^T A_T \Pi_T,$$

where Π_T is a projection $\mathbb{R}^n \to \mathbb{R}^3$ mapping the index positions i, j, and k of $\mathbb{R}^n$ to the three-dimensional space. The projections Π_T must be modified in the obvious way for triangles with constrained nodes in $\mathcal{N}_D$ such that the products with the corresponding basis functions do not appear in the global system.

If the algorithms do not need the stiffness matrix explicitly, but need the product of A^h with a vector, the assembly need not be performed, but the matrix can be stored in terms of its element matrices. Note that

$$A^h x = \sum_{T \in \mathcal{T}} \Pi_T^T A_T \Pi_T x. \tag{4.11}$$

Constrained nodes must be eliminated, because an explicit Dirichlet condition would violate the matrix symmetry. Multiplying A^h with a vector in this *distributive* representation is more expensive than using an assembled version of A^h, but it still may be the most effective data structure if the basic structure is *element-oriented,* as, for example, in the KASKADE code [49]; see also Deuflhard, Leinen, and Yserentant [35] and Roitzsch [82].

Exploiting symmetry. For self-adjoint systems, the stiffness matrix will be symmetric, $a_{i,j} = a_{j,i}$. Clearly, this can be exploited in the matrix-oriented and element-oriented representation by storing only those $a_{i,j}$ with $i \geq j$. In the edge-oriented representation, this is also trivial if we use a nonoriented graph of edges, each edge having two attributes for nonsymmetric and only one attribute for symmetric matrices.

If the edges are stored as a directed graph of *semi-edges* the data structure becomes more complicated, because the ordering of the nodes induced by the natural indexing in a matrix representation must be introduced artificially. An appropriate storage scheme in a sparse matrix data structure is introduced in George and Liu [39] and Schwarz [93]. Whether this compact representation of matrices is suitable depends on the algorithms used in the solution process. If only matrix-vector products are necessary, the compact representation is efficient. However, for successive relaxation algorithms the nonsymmetric representation is usually preferable, because it allows faster access to the matrix elements in the same row.

4.3. Special cases of finite element meshes

In this section we will discuss various adaptive mesh structures and techniques. We will start with the simplest structures and then introduce the more complicated ones.

4.3.1. Uniform meshes. In geometrically simple cases uniform meshes can be used. For a start, consider the unit square $\Omega = (0,1) \times (0,1)$. We introduce the uniform family of triangulations $(\mathcal{T}^h, \mathcal{N}^h, \mathcal{E}^h)$:

$$\begin{aligned}
\mathcal{N}^h &= \{(ih, jh) \mid i,j = 0,1,2,\ldots,N\},\\
\mathcal{E}^h &= \{\overline{(ih,jh)(ih+h,jh)} \mid i = 0,1,\ldots,N-1, j = 0,1,2,\ldots,N\} \cup\\
&\quad\ \{\overline{(ih,jh)(ih,jh+h)} \mid i = 0,1,\ldots,N, j = 0,1,2,\ldots,N-1\} \cup\\
&\quad\ \{\overline{(ih,jh)(ih+h,jh+h)} \mid i,j = 0,1,\ldots,N-1\},\\
\mathcal{T}^h &= \{\triangle((ih,jh)(ih+h,jh)(ih+h,jh+h)) \mid i,j = 1,2,\ldots,N-1\} \cup\\
&\quad\ \{\triangle((ih,jh)(ih,jh+h)(ih+h,jh+h)) \mid i,j = 1,2,\ldots,N-1\},
\end{aligned}$$

FIG. 4.3. *Generation of uniform meshes by affine transformations.*

where $h = 1/N$. This construction is called a *square mesh triangulation.* All triangles in $\mathcal{T}^h$ are congruent. The stiffness matrix derived for the Laplace operator on the square mesh triangulation is the same as that obtained for symmetric second order finite differences on the grid $\mathcal{N}^h$.

If the diagonal edges are dropped, we will obtain a triangulation with quadratic elements. For Laplace's operator, the finite element system for quadratic bilinear test and trial functions is equivalent to a discretization with the nine-point difference stencil (except the scaling)

$$\begin{bmatrix} 1 & 1 & 1 \\ 1 & -8 & 1 \\ 1 & 1 & 1 \end{bmatrix}.$$

General uniform meshes can be constructed from the square mesh triangulation by affine-linear transformations $M : (0,1)^2 \to \mathbb{R}^2$ and/or by selecting a subset of the triangles. For an example, see Fig. 4.3.

Uniform triangulations can be used only for *compatible* regions that can be obtained from the square mesh triangulation by affine mapping and selection of elements. Uniform meshes are characterized by the feature that the edges can have only three different directions and lengths. Correspondingly, such a mesh is also called *three-directional* (see Blum [16]).

On uniform meshes symmetry arguments show that certain error terms in the error expansion cancel, leading to *superconvergence effects.* Furthermore, the uniformity can be exploited to show the existence of *global* asymptotic error expansions; see Blum, Lin, and Rannacher [16]–[18].

On the other hand, for many practical applications, uniform meshes are not sufficient. Even for general polygonal domains, a correct representation of the boundary is not possible with globally uniform meshes. If, additionally, the solution has a different character in different parts of the domain, adaptivity becomes mandatory. The potential loss in performance and accuracy without adaptivity may be dramatic. With uniform meshes one will get too fine meshes where the solution is smooth, or too coarse meshes where small scale features must be resolved.

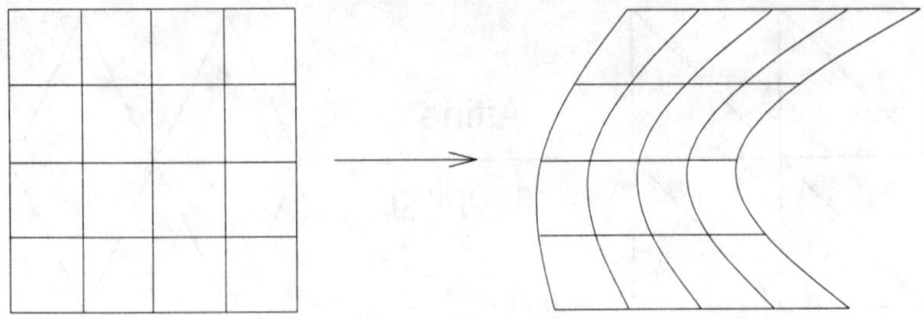

FIG. 4.4. *Quasi-uniform mesh.*

4.3.2. Quasi-uniform meshes. One possible way to generalize uniform meshes is to use *quasi-uniform* meshes. A quasi-uniform mesh is simply obtained by relaxing the requirement that the transformation $M : (0,1)^2 \to \Omega$ be affine. If more general mappings are used, we get quasi-uniform triangulations.

Depending on the context, the mapping M must satisfy certain regularity conditions. Curved boundaries can be accommodated because in the image space the elements have curved boundaries. An example is shown in Fig. 4.4, where M is a mapping that distorts the mesh nonlinearly. In general, however, the construction of a suitable mapping M may be as difficult as the solution of the original problem. The asymptotic error analysis of Blum, Lin, and Rannacher [16]–[18] remains valid only if the mapping M is sufficiently smooth.

Usually, the minimal requirement on M is that it be a homeomorphism (that is, a bijective, continuous mapping with continuous inverse), so that the quasi-uniform mesh is topologically equivalent to the square mesh triangulation.

A simple but useful construction of M consists of an affine mapping so that the square mesh triangulation completely covers Ω. Then the mapping is modified in the triangles that intersect $\partial\Omega$ by moving their corners to a point on $\partial\Omega$. The resulting mapping M can be defined as piecewise affine on the appropriate subsets of the square mesh triangulation. This construction is the equivalent of the Shortley–Weller discretization of general boundaries in the context of finite difference methods. For an example, see Fig. 4.5.

Using a homeomorphic mapping M that is piecewise affine on each triangle of the square mesh triangulation (and by selecting an appropriate subset of the triangles), more general meshes can be generated. Because of its topological equivalence to a subset of the square mesh triangulation, each interior node in $\mathcal{N}_I$ has exactly six neighbors.

For piecewise affine mappings, the system assembly can be performed, as suggested in equation (4.8), by integrating the shape functions. For nonlinear mappings the process must be generalized appropriately. The most important special cases are the so-called isoparametric elements.

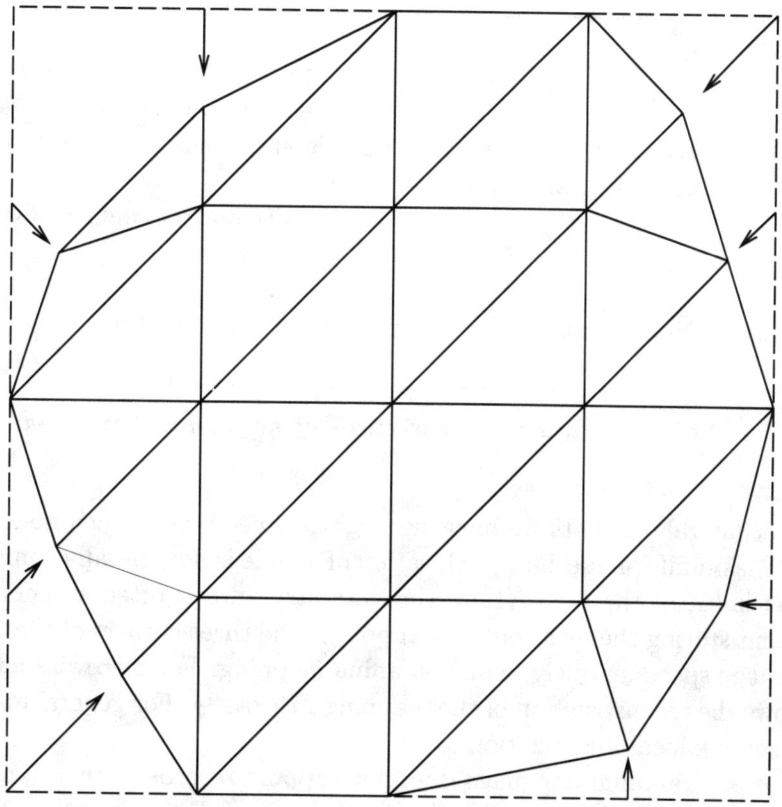

FIG. 4.5. *Uniform mesh with boundary modification.*

Data structures for quasi-uniform meshes. Uniform and quasi-uniform meshes can be represented in *arrays of nodes*. Each node is associated with an index position in a two-dimensional array. For quasi-uniform meshes, the mapping M must be stored additionally.

For uniform and quasi-uniform meshes derived from the square mesh triangulation we can use array structures, such that the access to individual nodes and sweeps through the nodes can be organized efficiently by index calculations and nested loops, respectively. The computational domain is called *logically rectangular*. The data structure for a piecewise affine, quasi-uniform mesh could be as given in Fig. 4.6, where we display an example data declaration in a C-like syntax. The generalization of this structure to more dimensions is straightforward.

The explicit representation of the diagonal could be avoided if all equations are scaled so that diag = 1. This is especially useful for saving repeated multiplications with diag^{-1}. For self-adjoint problems half of the off-diagonal elements could be saved if we explicitly use the symmetry of the matrix.

The components a1, ..., a6 for each node N_i are a representation of the attributed semi-edges $(\overline{N_i N_j}, a_{ij})$ with $\overline{N_i N_j} \in \mathcal{E}$ (see (4.9)), where the neighbors N_j are implicitly defined by the (quasi-)uniformity of the mesh.

```
struct Node {
        real u1, u2;    // approximate solution,
                        // several instances may be necessary
        real f, r;      // right-hand side and residual
        real diag, a1, a2, a3, a4, a5, a6;
                                        // matrix coefficients
}

struct Node mesh[N][N];   // mesh with N x N nodes
```

FIG. 4.6. *Data structures for piecewise affine quasi-uniform mesh.*

For quadrilateral elements we must use eight *connections* at each node.

For quasi-uniform meshes, the location of a node is not trivially computable from the index of the node. Here the data structure is often extended by a component storing the location of each point. The three corners of the triangle in the image space uniquely define an affine mapping. For piecewise affine M, this allows the reconstruction of the mapping M exactly. For general mappings it represents a local linearization.

If the programming language does not support records (structs), the data structure must be represented by either K arrays of dimension $N \times N$, or by an array of dimensions $K \times N \times N$, where K is the number of components in the record. The latter technique is only applicable if all components of the structure are of the same type.

Though these different representations (including all the hybrid forms imaginable) are logically equivalent, the different representations do affect the associated program organization. Additionally, they produce different memory layouts that may severely affect the performance of the corresponding program. This is particularly true for vector architectures with highly interleaved memory and superscalar systems with hierarchical memory structure (cache). These aspects are discussed in Bonk and Rüde [19].

Using elementary arrays often leads to more uniform access of the data so that vectorization becomes simpler, whereas the representation with structures (records) usually leads to better locality, because logically related data is stored in consecutive memory cells. This will have advantages on many modern, superscalar, and cache-based computer architectures.

The data organization and the access routines become more complicated if the region is derived from a *subset* of a square mesh triangulation. Typically, an organization by *grid lines* is used, where for each horizontal grid line the indices of the intersections with the boundary are stored. If the computational domain is convex, only one start- and one end-index are needed, but in general situations a variable number of index-pairs are required. Here the data structure becomes intricate. For an attempt to implement a general grid

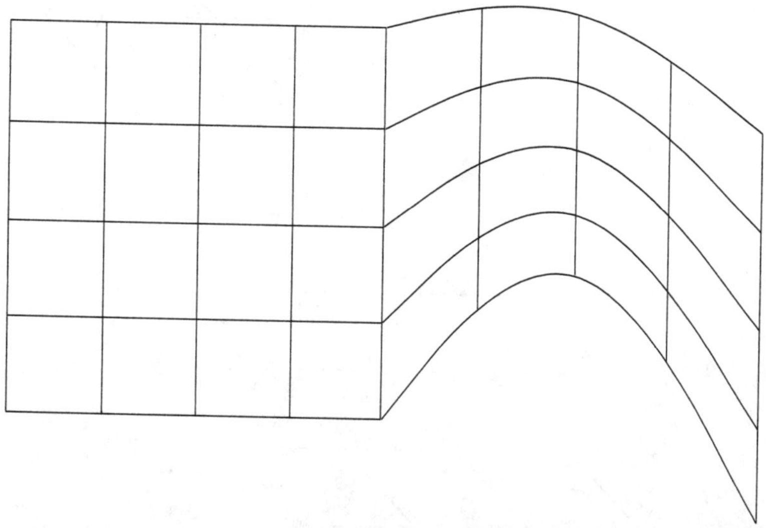

FIG. 4.7. *Piecewise quasi-uniform mesh.*

package based on these techniques, see Brandt and Ophir [31].

4.3.3. Piecewise uniform and quasi-uniform meshes. The basic idea of piecewise uniform meshes is straightforward. Several patches with (quasi-)uniform meshes are used to cover the domain. We distinguish between overlapping and nonoverlapping patches. As we have seen above, the most difficult aspect of the data structures are nonconvex computational domains. Piecewise (quasi-)uniform meshes may now be used to represent nonconvex domains with patches such that each patch is convex. Thus the data structure for each patch can be simple.

If the patches are nonoverlapping, they can be considered as macro-elements. Interface conditions describe their interaction with neighboring patches. Techniques related to this idea are *block-structured grids* (see Hempel and Ritzdorf [46]), *domain decomposition* (see Quarteroni [78] and the references therein), or the *substructuring technique*. These techniques describe solution algorithms suitable for piecewise (quasi-)uniform grids.

For the overall triangulation to be conforming, the nodes along the patch boundaries must be conforming across the interfaces. Thus single patches may not be defined or modified (refined) independently. A piecewise quasi-uniform mesh consisting of a uniform and a quasi-uniform part is shown in Fig. 4.7.

In practice, the condition that the mesh be conforming across the interfaces is often dropped, so that the patches can be refined independently. The interface conditions (for the nonconforming nodes) must be designed such that the solution has the required regularity (C^m). These interface conditions are often still difficult to treat in the solution algorithm; for example, they usually violate the symmetry of the discrete system.

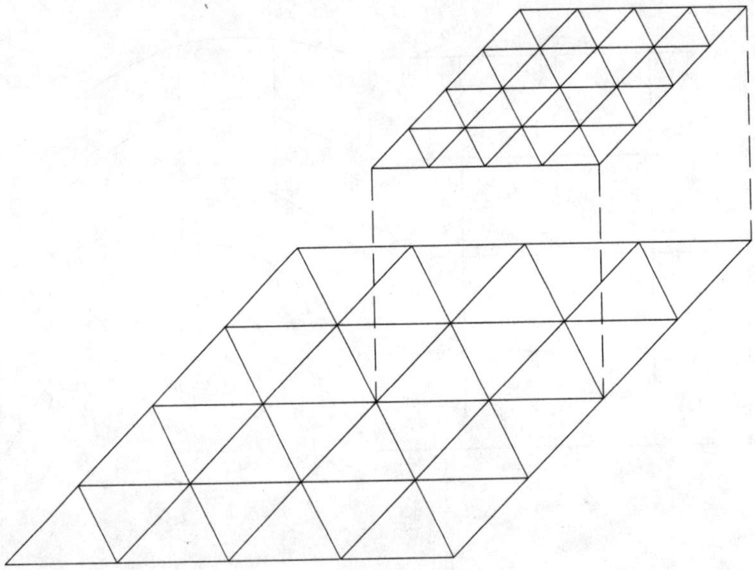

FIG. 4.8. *Two-level FAC mesh.*

The technique becomes more flexible if the patches are allowed to overlap; however, the definition of consistent interface conditions may become more complicated. Overlapping patches lead to another class of domain decomposition algorithms, and in particular to *iterative refinement* techniques (see Widlund [107]). This is also related to the *multilevel adaptive technique* (MLAT) of Brandt [27] and Bai and Brandt [6], and to the *fast adaptive composite grid technique* (FAC) (see McCormick and Thomas [62] and McCormick [57]; see also the multigrid methods with local refinement in Hackbusch [43]). For example, the FAC technique is usually applied with fully overlapping, uniform patches and nonconforming nodes; see Fig. 4.8. The correct treatment of the interface conditions is nontrivial and is discussed in McCormick [57].

For piecewise uniform meshes with refinement we mention the recent introduction of data structures with a high level interface that hides possible machine dependencies in parallel and distributed environments; see Lemke and Quinlan [51].

A quad-tree-based data structure for piecewise quasi-uniform, adaptively refined meshes in a multilevel context is discussed in Hemker, van der Maarel, and Everaars [45]; see also §4.5.

Piecewise uniform or quasi-uniform meshes can be implemented with the data structures already discussed. Additionally, the interface consistency must be maintained. This has a topological as well as an algebraic aspect. The triangulations must be conforming at the interfaces, or suitable interface conditions for the nonconforming nodes must be imposed. The algebraic processes for obtaining a solution of the global problem must suitably support the interface conditions. In typical domain decomposition algorithms and the FAC technique, the global stiffness matrix is only constructed implicitly and

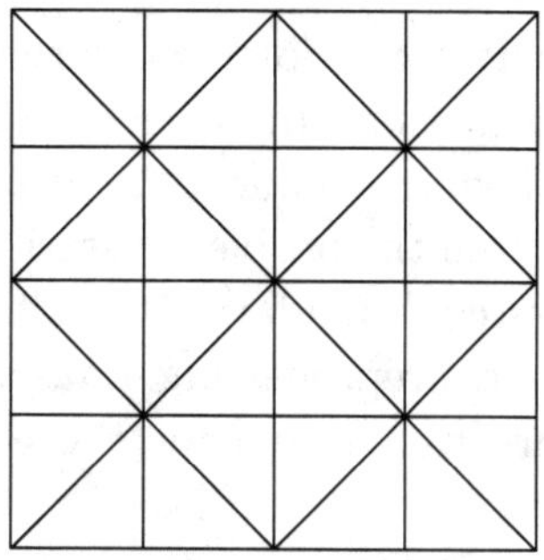

FIG. 4.9. *Mesh with four or eight semi-edges at each interior node.*

the global solution is enforced by alternating between inner iterations to solve the subproblems, and outer iterations to enforce global convergence, including the interface conditions.

4.3.4. Unstructured meshes. The meshes considered so far are called *structured meshes*. The nodes in the interior of a (quasi-)uniform patch with triangular elements all have six neighbors, and therefore the rows of the matrix (4.7) can be stored in seven variables; see the program fragment in Fig. 4.6.

Unstructured meshes sacrifice this property. A node in an unstructured mesh may have any number of neighbors. Thus any triangulation satisfying the conditions T1–T4 can be represented, and therefore the data structure must be extended so that an arbitrary number of off-diagonal elements can be stored. An example of an unstructured mesh is shown in Fig. 4.12. Unstructured meshes are the basis of our discussion of adaptive processes in §4.4.

Even unstructured meshes have the property that the *average* number of edges at a node is six. This may be exploited in the memory allocation routine if the number of six semi-edges is only exceeded at a small number of nodes. Then six semi-edges are provided as a default at each node, as is a dynamic list for additional semi-edges. The overflow treatment is not critical for performance if we assume that the number of nodes with overflow is small compared to n. This conforms nicely to the type of meshes generated by regular refinement.

A counterexample of a mesh where the number of semi-edges at interior nodes is either four or eight, so that only the average is six, is shown in Fig. 4.9. This mesh might have been constructed with a bisection-based refinement strategy.

4.4. Adaptive techniques

In the previous sections we have discussed mesh structures supporting various degrees of adaptivity. Now, we will discuss the mesh adaptation process itself and the features in the data structures necessary to support it.

4.4.1. A priori adaptivity. Some cases requiring adaptivity may still be handled by a priori information. In these situations the resolution needed in different parts of Ω is known in advance. An example is the case of singularities that can be anticipated from the problem description, like re-entrant corners or discontinuities in the source, boundary, or coefficient terms. Here an adaptive mesh can be created in a preprocessing phase. Quasi-uniform or piecewise quasi-uniform meshes are well suited for the a priori generation of adaptive meshes.

In the remainder of this paper we will assume that mesh adaptation in a preprocessing phase is not feasible, because the necessary information is not available a priori. In nonlinear problems, in particular, it may not be possible to construct a mesh without first having an approximate solution, but for simpler problems it is also desirable to have robust and user-friendly algorithms that can resolve local phenomena without requiring the user to explicitly provide an appropriately adapted mesh. A discussion comparing the *static* (a priori) and *dynamic* (self-)adaptive approach is given by Rivara [81].

4.4.2. Self-adaptive refinement. We will now discuss the aspects related to a dynamic approach, where the meshes are generated as a part of the solution process.

We assume that an *initial* triangulation $\mathcal{T}^{h_0}$ of the domain Ω is given, but that this initial triangulation is not yet sufficiently fine to obtain an accurate enough solution, and it is not yet sufficiently adapted to resolve local features of the solution well enough. The initial triangulation may be used to calculate a first approximate solution u^{h_0}, and based on this, a *mesh refinement* process may be started. Typically, this is related to a local *error estimator* or *error indicator*, telling where in the domain refinement would be computationally most profitable; see Babuška, Miller, and Vogelius [4], Bank and Weiser [9], and Deuflhard, Leinen, and Yserentant [35]. As an alternative, we recommend the error estimates of §3.3.

With the error estimate a new mesh $\mathcal{T}^{h_1}$ is constructed, the corresponding equations are solved, and, if necessary, the whole process is repeated; see the last diagram of Fig. 4.10. In general, a sequence of triangulations $\mathcal{T}^{h_0}, \mathcal{T}^{h_1}, \mathcal{T}^{h_2}, \ldots$ is constructed until sufficient accuracy is obtained or until the computing resources are depleted.

To simplify the notation, and to avoid multiple indices, we will use $\mathcal{T}^J$ instead of $\mathcal{T}^{h_J}$, and analogously for other identifiers in the remainder of this chapter.

4.4.3. Algorithmic requirements. Before we can discuss appropriate data structures, we need to analyze the mathematical properties and the algorithms that will be applied on the mesh hierarchy.

If we proceed in only one direction from coarser to finer grids with direct solvers, where the coarse grid solutions are only being used to define a new

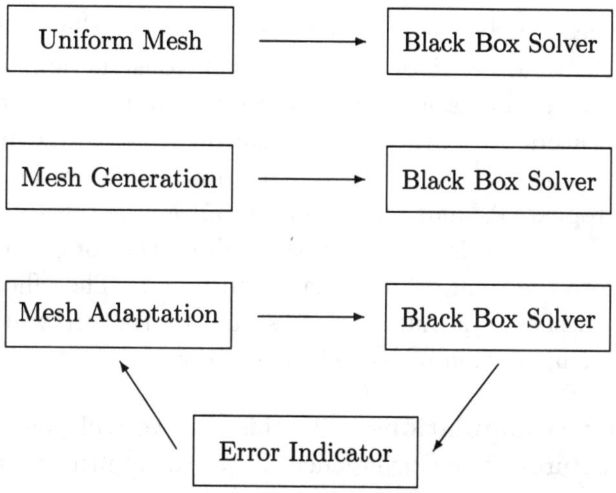

FIG. 4.10. *Classical solution strategies for elliptic problems.*

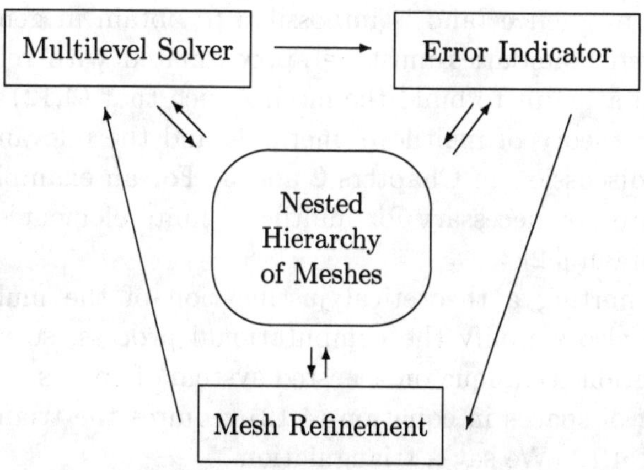

FIG. 4.11. *Integrated solution strategy for elliptic problems.*

mesh, then the relation between the meshes is of little importance.

However, for iterative solvers it is important that the information on $\mathcal{T}^J$ can be used to provide an initial guess for the solver on $\mathcal{T}^{J+1}$. A *prolongation* from $\mathcal{T}^J$ to $\mathcal{T}^{J+1}$ must be available. This means that an old grid must provide error estimates and an initial guess for the fine grid solution. Here there are still no restrictions on the new mesh, except that the interpolation operator should not be too difficult to compute. When the new grid has been created and initialized, the old grid may even be erased to save storage.

If we turn to multilevel solvers, this is different because the coarse meshes are required for processing the finer levels. Multilevel solvers are a particularly efficient class of iterative methods, as discussed in Chapters 2 and 3. These solvers fit nicely into the self-adaptive refinement scheme, because the sequence of refined meshes can be used as the basis for a multilevel iteration. This, however, poses several restrictions on the mesh construction.

In addition to prolongations from coarse to fine meshes, we now also need *restriction* operators to transport information from a fine mesh, $\mathcal{T}^J$, to a previous coarse one, $\mathcal{T}^{J-1}$. Within a full multilevel solver, the restriction must be used repeatedly and recursively on all levels $1 \leq l \leq J$ down to the coarsest. The coarser levels are needed throughout the computation and cannot be deleted before the final solution has been found. The efficient integration of the coarser meshes in a multilevel mesh data structure will be the basis of an efficient implementation of the solver; see Fig. 4.11.

4.4.4. Nested triangulations. To this end we will now discuss suitable mesh data structures. The convergence theory for multilevel iterations partly relies on a nested sequence of discrete spaces (see also equation (2.26)):

$$V_0^1 \subset V_0^2 \subset \cdots \subset V_0^J. \tag{4.12}$$

Though practical experience shows that this sequence of inclusions is not necessary for convergence (and is impossible to obtain in general situations, e.g., when curved boundaries must be approximated with refined triangular meshes), we will attempt to build the meshes such that (4.12) is satisfied. For the convergence theory of multilevel methods and the relevance of condition (4.12), see the discussion in Chapters 2 and 3. For an example showing that nested spaces are not necessary for multilevel finite element solvers, see also Löhner and Morgan [52].

Besides supporting a theoretical justification of the multilevel process, nested systems also simplify the computational process, so we will attempt to base our solution technique on a nested system of spaces.

The nesting of spaces in equation (4.12) requires the triangulations to be nested; see Fig. 4.12. We say a triangulation

$$(\mathcal{T}^J, \mathcal{N}^J, \mathcal{E}^J)$$

is nested in

$$(\mathcal{T}^{J-1}, \mathcal{N}^{J-1}, \mathcal{E}^{J-1})$$

if

$$\begin{aligned} &\mathcal{N}^{J-1} \subset \mathcal{N}^J, \\ &\bigcup_{E \in \mathcal{E}^{J-1}} E \subset \bigcup_{E \in \mathcal{E}^J} E, \\ &T^J \in \mathcal{T}^J \Rightarrow \exists T^{J-1} \in \mathcal{T}^{J-1} \quad \text{such that } T^J \subset T^{J-1}. \end{aligned} \tag{4.13}$$

Next we discuss techniques for creating nested, regular families of triangulations.

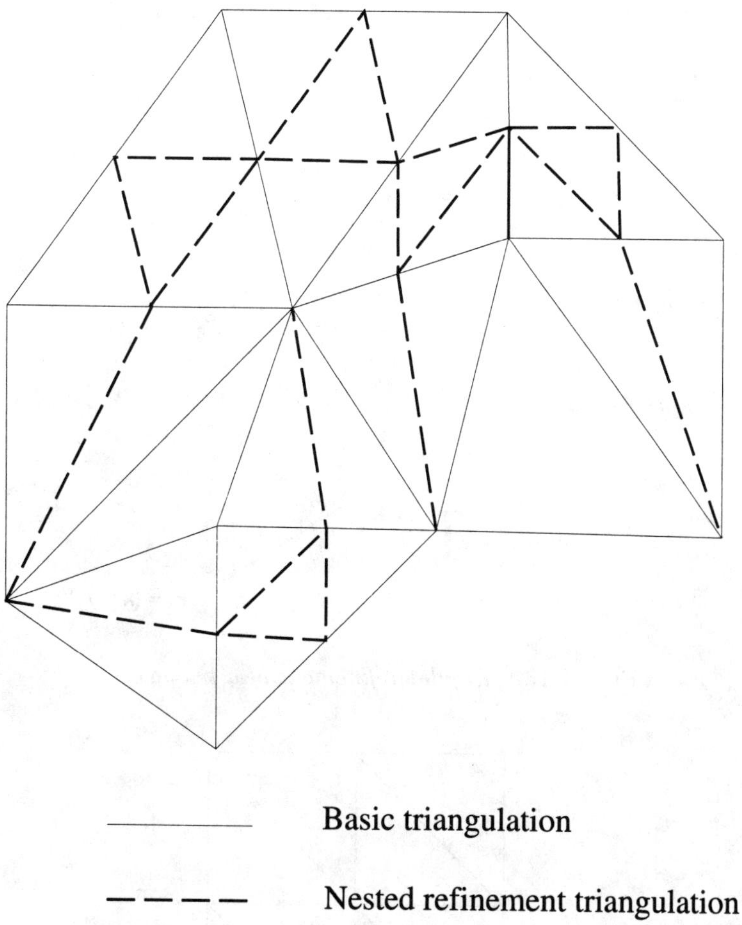

FIG. 4.12. *Nested triangulations.*

4.4.5. Regular refinement. The standard process for refining a given triangulation is *regular refinement.* The regular refinement of an individual triangle is shown in Fig. 4.13. It consists of introducing new nodes on the midpoint of all edges and connecting these midpoints in each triangle. Each original edge is split into two new edges of half the old length. Clearly, the new triangles T_1, T_2, T_3, and T_4 are all congruent and similar to the original triangle T.

A regularly refined mesh is obtained by regular refinement of all its triangles. It is well defined for any conforming triangulation and yields a new triangulation that is also conforming. The meshes are nested and so are the finite element spaces generated. Successive regular refinement generates a *regular family of triangulations* $\mathcal{T}^0, \mathcal{T}^1, \mathcal{T}^2, \ldots$ (see conditions T5 and T6 in §4.2), because $h(\mathcal{T}^i) \longrightarrow 0$ and $\sigma(\mathcal{T}^i) = \text{const}$.

Also note that each interior node generated on the midpoint of an edge has exactly six neighbors, and those which are coinciding with the nodes of the old

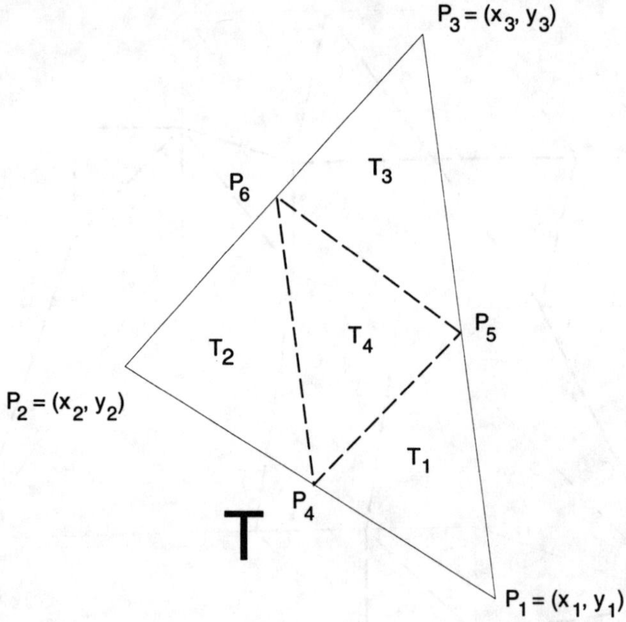

FIG. 4.13. *Regular refinement of a triangle.*

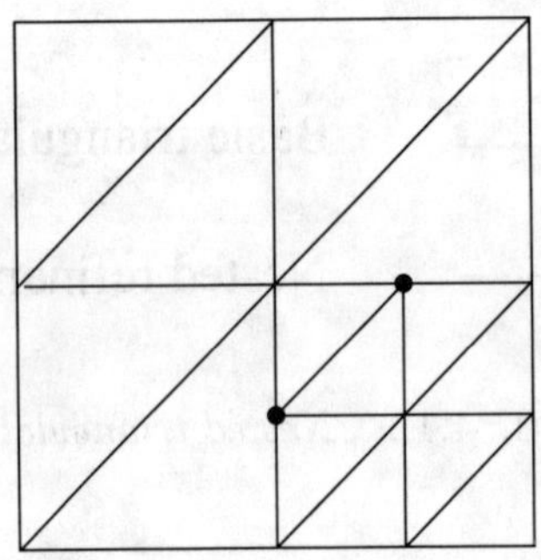

FIG. 4.14. *Regularly refined mesh with nonconforming nodes.*

triangulation inherit the neighbor structure from the coarse mesh. Augmented regular refinement is the basis of the adaptive codes of Bank [10] (PLTMG) and Leinen [49] (KASKADE).

Clearly, however, regular refinement alone cannot support adaptivity if *conforming* triangulations are required. Triangles excluded from regular refinement will have nonconforming nodes if an adjacent triangle has been refined. This is illustrated in Fig. 4.14.

There are two options. As for overlapping uniform grids, we can simply accept the occurrence of nonconforming nodes, which will require us to treat them as in the FAC method. The resulting structures will effectively be

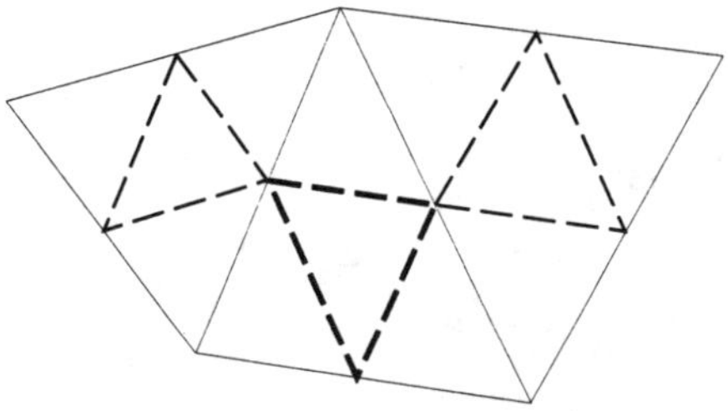

FIG. 4.15. *Regular refinement induced by two nonconforming nodes on different sides.*

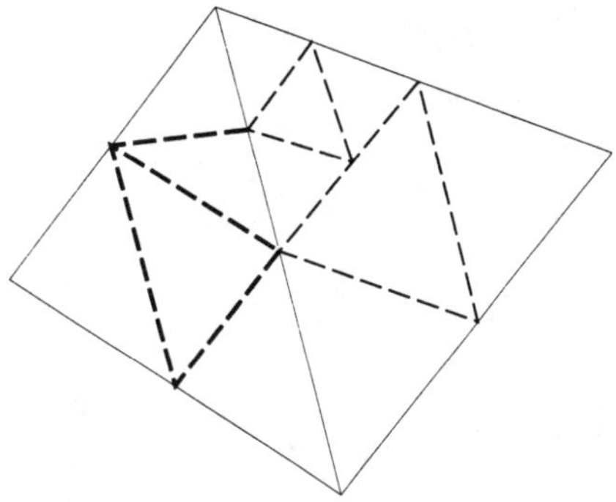

FIG. 4.16. *Regular refinement induced by two nonconforming nodes on the same side.*

similar to FAC, with the difference that now the basic mesh is unstructured. Alternatively, we can attempt to build conforming nested meshes.

4.4.6. Refinement by bisection. Even if the basic refinement uses regular division, we need bisection to make the meshes conforming. Figure 4.12 shows a coarse mesh where five triangles are being refined regularly. This is made conforming by four additional triangles that are bisected. The PLTMG and KASKADE programs both use bisection for the *closure* of a triangulation. The bisecting edges that make the triangulation conforming are called *green edges.* Green edges are removed before still finer meshes are constructed. This avoids one potential problem that can arise with bisection: Angles are halved, and if this happens repeatedly, then the regularity condition T6 in §4.2 may be violated.

Note that additional problems can arise in triangles when two edges both have a nonconforming node (see Fig. 4.15) or when one edge has two or more nonconforming nodes (see Fig. 4.16). Refinement algorithms avoid these situations by performing one more step of regular refinement in these triangles, as shown by the thin dotted lines in Figs. 4.15 and 4.16. This may cause new nonconforming nodes that must be treated analogously.

Finally, note that augmented regular refinement, as described above, will *not* produce strictly nested triangulations. Triangles where a green edge has been removed for a regular refinement will not have nested triangulations, and therefore the spaces associated with these triangulations will not be nested either. The experimental results of Bank [10] and Leinen [49] suggest that this does not severely affect convergence, but it leaves a gap in the associated theory.

An alternative way to generate regular families of triangulations is by using bisection in the first place. This avoids the trouble with green refinement and will produce triangulations that are strictly nested. In bisection algorithms,

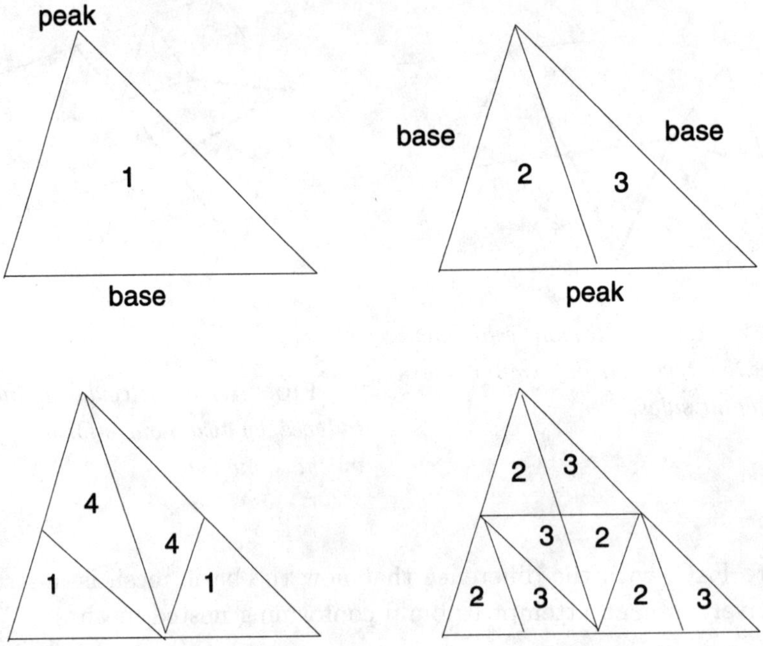

FIG. 4.17. *Four similarity classes generated by newest node bisection.*

care must be taken that the minimum angle condition not be violated. To avoid nonconforming nodes, triangles must be bisected in pairs. The minimum angle condition is enforced in algorithms based on the usual bisection by either requiring that no angle be bisected more than once or by bisecting only the angle opposite the longest edge of a triangle. These strategies are used by Rivara [81] and Sewell [95]. Different bisection strategies are being compared to regular refinement by Mitchell [64], [65]. For related results and the extension to the three-dimensional case, see Bänsch [11].

The *longest edge bisection* algorithm of Rivara [80] first bisects all triangles selected by the error indicator by connecting the midpoint of the longest edge to the opposite vertex. Now triangles with a nonconforming node are bisected. This may create new incompatibilities, so the process must be repeated until all nodes are conforming. Rivara has shown that this longest edge bisection will terminate after a finite number of steps.

Alternatively, we can use a *newest node bisection* algorithm. The newest node is called the *peak* of a triangle and the opposite side, the *base* of the triangle. After dividing a triangle by connecting the peak to the midpoint of the base, the new node becomes the peak of each of the two new triangles. Sewell [95] has shown that all the triangles created in this form belong to four similarity classes, so that the minimal angle condition is guaranteed. The algorithm and the similarity classes are illustrated in Fig. 4.17.

4.4.7. Self-adaptive refinement with uniform patches. Our discussion of self-adaptive refinement algorithms has so far been based on the individual refinement of elements in an unstructured mesh. It should be mentioned that there are attempts to construct self-adaptive codes based on piecewise uniform meshes. In these algorithms the information by the error estimator is postprocessed to find simply structured patches that contain the region to be refined. The goal of these algorithms is to maintain the computationally simpler structure of piecewise uniform patches, usually at the price of refining in somewhat too large regions.

4.4.8. Virtual global grid refinement. The idea of *virtual global refinement* is linked to the concept of *fully adaptive multigrid methods* (FAMe), as introduced in Rüde [88], [91]; see also Chapter 3. This refinement technique is directly related to multilevel solvers and hierarchical basis and leads to an algorithm where iterative solver, error estimate, and refinement are integrated into a single unit.

The idea is to use global regular refinement, so that each grid in principle covers the full domain. To accommodate adaptivity, *ghost nodes* are introduced. These nodes need not be allocated in the memory nor are there any computations performed with them. The key for this technique is to recognize that the ghost nodes do have associated numerical values defined by interpolation from the coarser grid, without needing any storage or processing. This interpolation may have to be applied recursively, if there are also ghost nodes on the coarser grid.

If the value of a node is defined through interpolation, this can be interpreted such that its *hierarchically transformed* value (*hierarchical displacement*) vanishes. In the notation of hierarchical bases, a ghost node is interpreted to have a value equal to zero. The technique of ghost nodes is therefore analogous to sparse matrix techniques where a rectangular matrix is stored in a compact form, exploiting the fact that many of its entries vanish. Here the same concept is applied for the meshes.

The sequence of global refinements forms an infinite recursive data structure. The goal will be to define a suitable *lazy evaluation* strategy. Variational multilevel algorithms can be reformulated so that nodes with vanishing hierarchical displacement never appear in the real computations. The resulting refinement method is algebraically equivalent to the FAC treatment of patch interface conditions; see Rüde [87].

A mechanism must be devised to convert a ghost node to a *live node*, if necessary, to achieve the accuracy of the computation. This is accomplished by the *adaptive relaxation* algorithm described in Chapter 3 and in Rüde [91]. In Chapter 3 it was shown that a suitable error indicator is given by the changes to a node during relaxation. These changes are just the current scaled residuals that occur in the error estimates of Corollaries 3.3.2 and 3.3.3. Relaxation must be performed in a multilevel algorithm anyway, so error estimation and solution go hand in hand.

The processing of live nodes that have ghosts among their neighbors requires access to the ghost node. To support this, we demand that live nodes never be exposed to real ghosts, and we introduce *guard nodes* instead. These are nodes whose information can be computed by interpolation from coarser levels. The guards are real data structures in memory that can be accessed to perform relaxation and other processing on neighboring nodes. Guards fence in the live nodes and have a function similar to the interface nodes on patch boundaries in the FAC method (see McCormick [57]). Virtual global grids can be interpreted as a FAC-like technique, where the form of a patch is not predetermined, but where *patches* can be generated dynamically on a node by node basis. The guards are nonconforming nodes in the finite element mesh.

The *guard nodes* also carry the burden of adaptivity, because they can be triggered to convert to live nodes when demanded by the prescribed solution accuracy. A guard that becomes a live node will generate new guards to keep the interface consistent, if necessary. Figure 3.19 shows the *live* nodes for a uniform virtual global grid representation of the function in Fig. 3.18.

Selecting the live nodes on a mesh level is equivalent to selecting a subset of the nodal basis (the live basis) of the corresponding mesh:

$$\mathcal{L}^k \subset \{B_i^k | N_i^k \in \mathcal{N}^k\}.$$

When we unite these bases,

$$\mathcal{L} = \bigcup_{k=1}^{J} \mathcal{L}^k,$$

we obtain a *generating system* for the virtual global grid space,

$$V_{\text{virtual}} = \text{span } \mathcal{L}.$$

$\mathcal{L}$ usually contains linearly dependent elements that must be eliminated if we want to create a basis. A basis can be obtained by omitting those basis functions in $\mathcal{L}$ that would not appear in the corresponding global grid hierarchical basis. For consistency, we must require that the supports of the live bases be nested:

$$\text{supp } \mathcal{L}^k \subset \text{supp } \mathcal{L}^{k-1}.$$

These definitions show that the virtual global grids define discrete spaces suitable for finite element approximation and the associated theory. Algorithms that perform finite element calculations on virtual global grids are discussed in Chapters 2 and 3 and in Rüde [91].

It should be understood that the virtual global grid technique is a concept that can be used with any of the classical mesh structures presented in §4.3. The implementation of virtual global grids using collections of live and guard nodes is just one alternative. A completely different idea is to use *hash-indexing* to access the live nodes of a virtual global grid. This is particularly attractive if the basic grid structure is simple, as for uniform or quasi-uniform grids.

4.5. Data structures for hierarchical finite element meshes

4.5.1. Relations in the multilevel hierarchy. We now discuss data structures for representing a hierarchy of finite element meshes. We again concentrate on the model case of two-dimensional triangular meshes, though many of the techniques may be applied for quadrilateral and other types of elements in a similar fashion. The purpose of this section is to extend the basic system $(\mathcal{T}^J, \mathcal{N}^J, \mathcal{E}^J, \mathcal{C}^J)$ to reflect the multilevel structure generated by self-adaptive refinement.

Elements on refined levels $T^J \in \mathcal{T}^J$ are contained in coarser triangles $T^{J-1} \in \mathcal{T}^{J-1}$. The inclusion of fine triangles in a coarse triangle generates a basic relation on triangles on different levels. This *"is-contained-in"* relation $\subset \mathcal{T}^J \times \mathcal{T}^{J-1}$ plays an important role in the element-based data structures of §4.5.2.

Fine mesh *edges* are also related to the coarse mesh. A coarse mesh edge may be bisected into two fine mesh edges so that a *"bisects-edge"* relation $\subset \mathcal{E}^{J-1} \times \mathcal{E}^J$ can be defined. Other fine mesh edges are cutting through a coarse mesh triangle to which they are then related by a *"bisects-element"* relation $\subset \mathcal{E}^J \times \mathcal{T}^{J-1}$. Still another relation can be defined to relate the edges in a regular refinement to the coarse mesh triangle.

The *nodes* of a refined mesh are either nodes at the same location as a node of the coarse mesh or they are the bisectors of a coarse level edge. In §4.5.3 the nodes on different levels will be considered distinct entities. If this is the case we can define a relation *"is-parent"* $\subset \mathcal{N}^{J-1} \times \mathcal{N}^J$ for nodes that occupy the same geometric location on two consecutive levels, and a relation *"interpolates"* $\subset \mathcal{N}^{J-1} \times \mathcal{N}^J$ to describe the interpolation dependency according to the bisection of a coarse mesh edge. This last relation must be generalized if more than just linear interpolation is used.

4.5.2. Element-based data structures. The inclusion relation for triangles between different levels defines a tree structure. Depending on the type of refinement, each triangle in $\mathcal{T}^{J-1}$ has either two descendants (bisection) or four descendants (regular refinement). Unrefined triangles become the leaves in the tree. An example is shown in Fig. 4.18.

Alternatively, we can extend the tree by giving the unrefined coarse triangles one single descendant identical to the parent triangle itself. This leads to trees where a finite element mesh is defined by a level in the tree. Note that here a mesh does not uniquely define the refinement tree through which it has been derived. At any level, dummy refinements can be added and it would be possible to have a refinement tree where only one additional triangle is partitioned from level to level. A multilevel solution process is naturally defined by the tree structure and will be inefficient when the refinement proceeds that slowly.

In any case, the tree is suitable as the frame data structure for the multilevel finite element structure, if it is augmented by additional information. Following

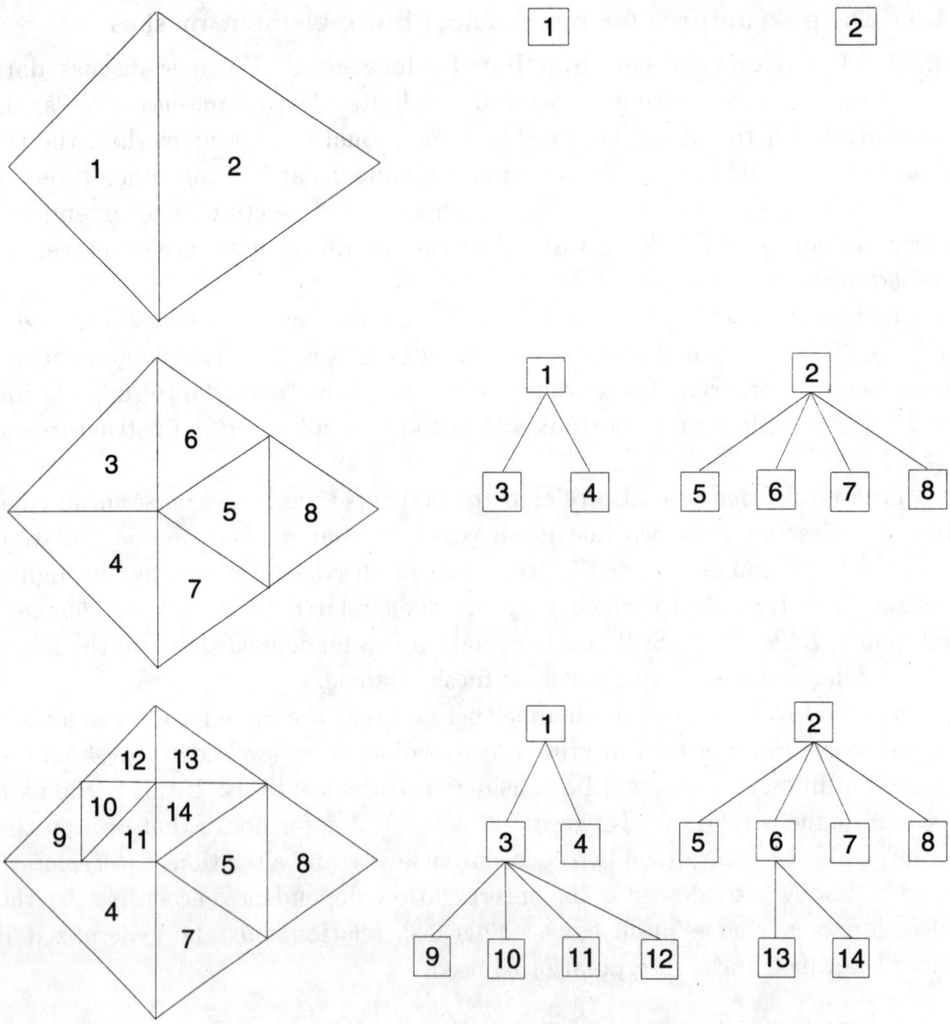

FIG. 4.18. *Construction of a* triangle tree *by successive refinement.*

Leinen [49] the element matrices can be stored with the triangles. These matrices implicitly contain the algebraic connections $\mathcal{C}$ between nodes (see equation (4.10)). Then the stiffness matrix is never assembled explicitly, but only implicitly, when matrix-vector products are formed (see equation (4.11)).

The nodes and edges are separate data structures and are referenced from the triangles in the tree. In Leinen's original KASKADE data structure, nodes are unique entities not assigned to any level. Since the solution values are stored as part of the nodes, this representation is primarily suited to the hierarchical basis algorithm that does not need individual representations of the solution on each level.

For algorithms that need several instances of the solution, like the BPX algorithm (see Bramble, Pasciak, and Xu [25]), the concept must be extended by providing a *stack* of values for each node.

The typical access routine for the element-based data structure is a recursion in the triangle tree, from where nodes and edges are referenced by pointers. For a detailed discussion, see Leinen [49].

Another element-based data structure is described in Hemker, van der Maarel, and Everaars [45]. Their data structure is based on quadrilateral *cells*, so the inclusion relation generates a so-called *quad-tree*. The BASIS data structure uses this tree of cells as the frame data structure that is augmented with geometric and algebraic information. Again, the access structure is a recursion through the quad-tree, from where all other information is referenced. Each cell is associated uniquely with its southwest corner, so that at the east and north boundaries special cells are required that only represent a piece of the boundary, not a full element. The data structure, however, has the advantage that the cells correspond uniquely to the nodes. At the boundaries of refined regions, nonconforming nodes, so-called *green walls*, are introduced. The BASIS data structure is used with cell-centered discretizations.

The structure generated by this is similar to refinement by piecewise uniform patches, as used in the MLAT and FAC techniques (see §4.3.3); however, the regions of refinement can be determined with more flexibility. The BASIS data structure therefore has some of the features of the *virtual global grids.*

The BASIS data structure also provides routines to *scan* through all patches of the present mesh. The action to be performed is specified as a subroutine parameter to the scanning routine. This idea can be interpreted as an implementation of the *iterator* technique introduced in §4.6 (see the discussion there).

4.5.3. Node-based data structures. In the element-oriented data structure of §4.5.2, a node is uniquely associated with a geometric location. Within multilevel algorithms, each node is attributed with solution values. In a general multilevel algorithm each level needs a separate set of solution instances, so we modify our notion of a node by assigning it to levels. Thus different nodes may occupy the same geometric location if they are on different levels of the mesh hierarchy.

For these extended nodes we introduce the *"is-parent"* relation between nodes on consecutive levels ($\mathcal{N}^{J-1} \times \mathcal{N}^{J}$). Together with the neighbor-relation $\mathcal{E}^J$ within a level (plus the connections $\mathcal{C}^J$ if general elements are used), this is a suitable frame data structure for the finite element hierarchy.

This is illustrated in Fig. 4.19, where the *"is-parent"* relation is displayed by thick dotted arrows. In contrast to the element-oriented approach this data structure is now *node-oriented.* The *node* (*vertex*) is the primary data type. *Edges* are stored separately.

In this data structure, there are no restrictions on the number of the connections between nodes or their topology. We can represent a general graph, and thus the physical interpretation ranges from one-dimensional next-

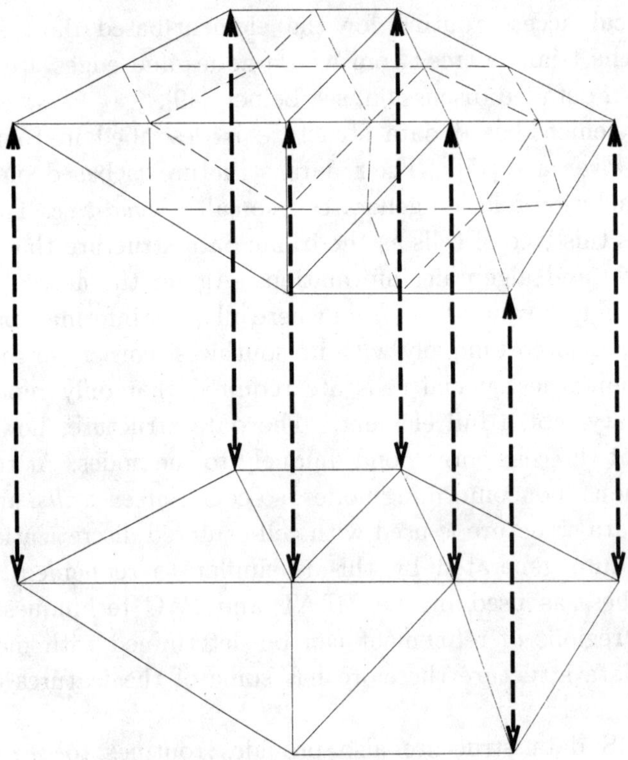

FIG. 4.19. *Hierarchy of two grid levels.*

neighbor interactions to general d-dimensional problems with $n \times n$ interactions.

Note that for a triangular mesh the information contained in both the element- and node-oriented data structures is the same. However, while the element-oriented approach supports the efficient execution of operations that need the element structure directly, these operations are expensive in a pure node-oriented structure. The elements must be constructed by finding the 3-cycles in the edge-graph. To be implemented with better efficiency, the node-oriented data structure must be augmented by additional information, e.g., by explicitly storing the relation *has-vertices* that gives explicit access to the elements.

On the other hand, the node-oriented structure contains an efficient sparse matrix representation if attributed edges (more generally, connections) are used. Thus matrix operations can be performed more efficiently than in the element-oriented structure unless the element-oriented structure is extended to store an assembled version of the stiffness matrix.

The node-oriented approach can be used to store the *algebraic structure* of any linear finite element problem, because it is equivalent to general sparse matrices. For example, tetrahedral meshes in three dimensions need only nodes and edges connecting them. Though the topology of the connections is more complicated than for triangles in two dimensions, the algebraic structure can be represented with attributed edges and nodes only. If we generalize the edges to connections, the data structure can also be used for rectangular, bilinear elements. These elements cause interactions between vertices along both diagonals of the element. These diagonals must become connections in the data structure, though they are not edges.

For complicated elements it may be necessary to have several variables associated with one node location. This can be dealt with in two alternative implementations. We may allow several nodes at the same geometric location, each node still having only one numerical value associated with it. A better approach, perhaps, is to introduce nodes with several degrees of freedom. Then, also, the edge (or connection) data structure must be extended to contain submatrices of the stiffness matrix instead of scalar coefficients. Thus, within a single level, a geometric location can only be occupied by one node.[2]

Note that the element-oriented approach is more tailored towards the special case with triangular elements. This makes the approach less flexible, but it allows the more efficient implementation of operations that are directly related to the form of the elements. This includes refinement algorithms and integrating the element matrices.

In a multilevel algorithm we must implement algorithmic components like *smoothing*, *calculation of the residual*, *calculation of the τ-term*, *coarsening*, and *interpolation*; see Chapter 3. If the multilevel algorithm is implemented with hierarchical transformations (see, e.g., Rüde [87]), then these operations are partially replaced by *calculation of the hierarchical transformation* and *reversing the hierarchical transformation.*

In any case, these operations require the execution of a nested loop of the form

> for all nodes $N \in \mathcal{N}^J$
>   for all connections (N, N', a)
>   perform operation with node N using N' and a

Our data structure must therefore guarantee efficient access to all nodes of a level, and for each node, efficient access to all edges (connections) and neighbors. Other operations in the multilevel process require the transfer of information between levels. The levels of the hierarchy are now strictly separated, so that each level is represented by its own finite element mesh. A typical basic mesh consisting of eight elements is shown in Fig. 4.19. The finer mesh is created by *regular* refinement of all triangles of the coarser mesh.

Adaptive multilevel methods on unstructured grids are studied by several authors. For further information, we refer to the work by Bank [10], Bastian [12], Leinen [49], Mitchell [64], and Rivara [81].

[2]This is only violated in situations like the slit domain.

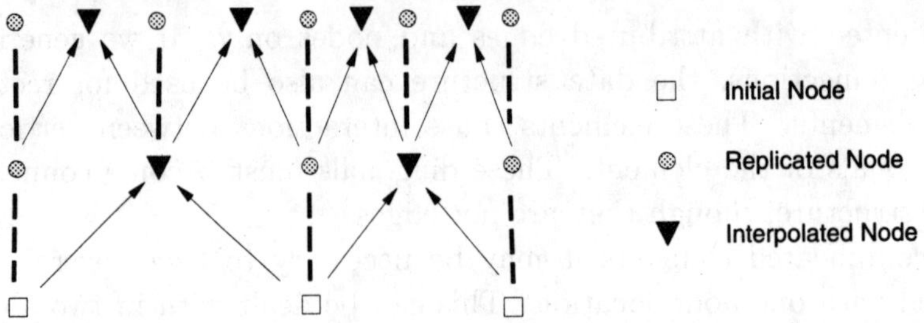

FIG. 4.20. *Different types of nodes.*

4.5.4. Classification of nodes. A closer analysis of the node-oriented structure shows that different nodes perform different tasks. For an efficient implementation of multilevel algorithms we must distinguish between these various types of nodes. This classification will be the basis of an *object-oriented design*, discussed in §4.6.

Each node is related to nodes on finer and on coarser levels. The nodes of the initial triangulation $\mathcal{N}^0$ are the only ones not being related to coarser ones. This requires a special type of *initial node* that can perform the functions typical for the coarsest level. For example, the initial nodes must support direct solution algorithms.

All other nodes are *refining nodes.* Two types of relations between the levels exist. A node may be a "copy" of a coarse node, meaning it occupies the same geometric location. We will call these nodes *replicated* nodes. The other nodes are related to the coarser grid by an interpolation procedure. These nodes are called *interpolated* nodes.[3]

Each replicated node has a *parent node.* This is a node with the same coordinates as the given one on the next coarser level as defined by the *is-parent* relation. Following the chain of parents to the coarsest possible level until we find either an initial or an interpolated node, we define the *ancestor node.* The ancestor node of an initial or interpolated node is defined as the node itself. The parent and ancestor relations play an important role in the implementation of the algorithm. The structure is illustrated in Fig. 4.20.

For replicated nodes we can use the fact that their behavior is partly defined by their parent and ancestor nodes. This can be exploited by not storing all information in the replicated node, but by implementing the operations based on references to coarser levels. The most obvious such information is the geometric location of a replicated node that is by definition the same as for its parent and its ancestor.

Depending on the solution algorithm, certain intermediate values of the solution process need not be stored in a replicated node, but may be accessed through the ancestor. This is especially convenient because in addition to saving storage, it eliminates the need to transfer these data items between

[3]The *interpolated nodes* correspond to the F-nodes, the *replicated nodes* to the C-nodes; see §3.6.

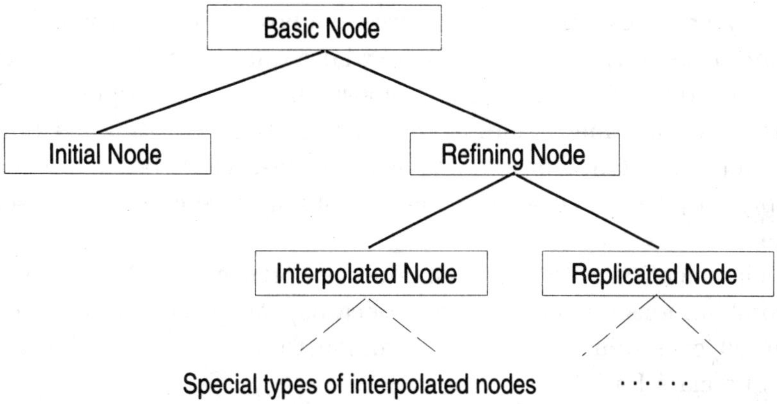

FIG. 4.21. *Class hierarchy for node data type.*

levels. Part of the coarsening, namely the transfer of the present approximate solution value to coarser levels, need not be performed explicitly. Similarly, values need not be transferred from coarse to fine meshes. Each replicated node, however, must store its own neighbor structure, because these relations depend on the mesh level.

Interpolated nodes have additional knowledge of how to obtain an interpolated value from the coarser grid, and how to perform a (local) hierarchical transformation. They are capable of storing the transformed value and the hierarchical displacement. For this, interpolated nodes need references to the coarse nodes that are used in interpolation. As remarked above, this can be implemented either directly, by referencing the coarse grid nodes from where the interpolant is calculated, or indirectly, by referencing the coarse grid edge that is bisected by the interpolating node. Both techniques must be generalized appropriately for nontriangular elements.

For quadrilateral elements an extra subclass of interpolated nodes will be needed representing the nodes in the center of a coarse element. Different types of interpolation may be implemented by different subclasses of interpolated nodes. Operations like the calculation of the τ-term must be implemented differently for replicated, interpolated, and initial nodes. The discussion so far results in node data types as shown in Fig. 4.21.

Assuming a refinement technique where new nodes are introduced at the midpoints of the edges, one may attempt to avoid storing the location of all except the initial nodes. For replicated and interpolated nodes the location is defined through the initial grid alone. To obtain the location of the nodes, however, it will be necessary to look up values from coarser meshes recursively. A quick calculation shows that calculating the locations of all nodes of one level costs worse than linearly in the number of nodes. This is unacceptable, so this approach is not feasible and we will have to explicitly store the location for all (or at least some) of the interpolating nodes.

Another distinction can be made between *free* and *constrained nodes. Free nodes* ($\mathcal{N}_I$) are those that become free unknowns in the solution process; *constrained nodes* ($\mathcal{N}_D$) include all those whose value is determined by Dirichlet boundary conditions or other direct constraints. An interpolated Dirichlet node may need a somewhat different interpolation when curved boundaries must be approximated. For systems of equations, where boundary conditions may apply only to one of several components, a more complicated setup will be needed.

The classification criteria for the nodes are orthogonal. A node may be any of the combinations replicated/free, replicated/constrained, interpolated/free, interpolated/constrained. Features common to all types of nodes are implemented in a class BasicNode.

4.5.5. (Semi-)edges and connections. The edges of the finite element graph are implemented as a separate data structure. They also serve as an implementation of the *"is-neighbor"* relation between nodes.

For triangular elements, the edge data structure can also be used to store the matrix coefficients. To enhance the performance, even symmetric matrices should be stored by using two *semi-edges.* Each node is associated with a list of *semi-edges* (or more generally, connections) that implement a directed, attributed edge in the mesh. Each semi-edge contains a reference to the neighboring edge and a numerical value.

To define a proper finite element triangulation, semi-edges must always exist in pairs: If node a has a semi-edge to node b, node b must have a semi-edge to a. Splitting an edge into two semi-edges enhances the efficiency of relaxation and residual calculation, because they allow fast access to the neighbors of a node. They automatically allow the representation of non-self-adjoint equations with the corresponding nonsymmetric systems.

As discussed above, the common treatment of geometric information (edges) and numerical information (matrix coefficients) must be generalized when the elements are more complicated. Here the numerical information is stored in a separate *connection* data structure replacing the semi-edges. The connections store the off-diagonal elements in the stiffness matrix and thus constitute a compact sparse matrix representation. For geometric operations, the edges must be stored additionally.

For higher order elements or for systems, a node may store several variables so that the connection data structure must be further generalized to contain not only a scalar coefficient, but a submatrix of the stiffness matrix. To keep the code extensible, the connection should therefore not only contain a simple real component to represent a scalar matrix entry, but it should use a user-defined *connection-value* data type that can be defined as an appropriate submatrix type if needed. For scalar equations this *connection-value* is of course of scalar type. Analogously, the diagonal matrix entry stored in the node data structure should be a general data type that can be extended from a simple scalar component to a matrix, if necessary.

4.5.6. Partially refined meshes. With these structures, we may implement a nested hierarchy of finite element meshes. The basic assumption so far was that on each level we construct and represent a full triangulation of the domain Ω. For local refinement, interpolated nodes are used to reduce the mesh spacing locally. Globally, however, we must copy all coarse grid nodes to become replicated nodes of the new mesh. If the region of refinement is small compared to the total domain, this is inefficient. Even with the data structures introduced above, we will waste storage, and in cases where the number of interpolated nodes of a level grows too slowly, even the optimal complexity orders in storage and processing would be lost.

As this is intolerable, typical multilevel algorithms deal with this by dropping the requirement that on each refinement level the mesh covers all Ω. We modify the construction such that a mesh on a particular level does not necessarily cover the whole domain.

The meshes on each level are simply restricted to domains where true refinement has been performed, that is, where interpolated nodes have been introduced. The refinement region of each level thus covers a subdomain Ω^J. These subdomains form a nested sequence

$$\Omega^J \subset \Omega^{J-1} \subset \cdots \subset \Omega^1 \subset \Omega^0 = \Omega.$$

On level J we do not store all nodes in $\mathcal{N}^J$, but only a subset

$$\tilde{\mathcal{N}}^J = \{N \in \mathcal{N}^J | N \in \bar{\Omega}^J\},$$

and similarly we only store the edges

$$\tilde{\mathcal{E}}^J = \{E \in \mathcal{E}^J | E \in \bar{\Omega}^J\}.$$

The real finite element mesh of level J must therefore be composed recursively by taking the lower level mesh outside the refinement region and the actual level mesh inside the refinement region:

$$\mathcal{N}^J = \tilde{\mathcal{N}}^J \cup (\mathcal{N}^{J-1} \setminus \Omega^J)$$

for the nodes and

$$\mathcal{E}^J = \tilde{\mathcal{E}}^J \cup (\mathcal{E}^{J-1} \setminus \Omega^J)$$

for the edges. To make this well defined, the boundary of any refinement region $\partial\Omega^J$ must consist of edges in the coarser triangulation $\mathcal{E}^{J-1}$. Furthermore, to keep the finite element meshes conforming, no interpolated nodes must lie on the boundary of the refinement region (see condition T4 of §4.2). Note that this cannot be accomplished by exclusively using quadrilateral elements. An example is shown in Fig. 4.22.

This complicated construction of the finite element meshes is not only motivated by storage considerations. It is also well suited for multi-level algorithms, which typically do not access all nodes $\mathcal{N}^J$ or edges $\mathcal{E}^J$ of a

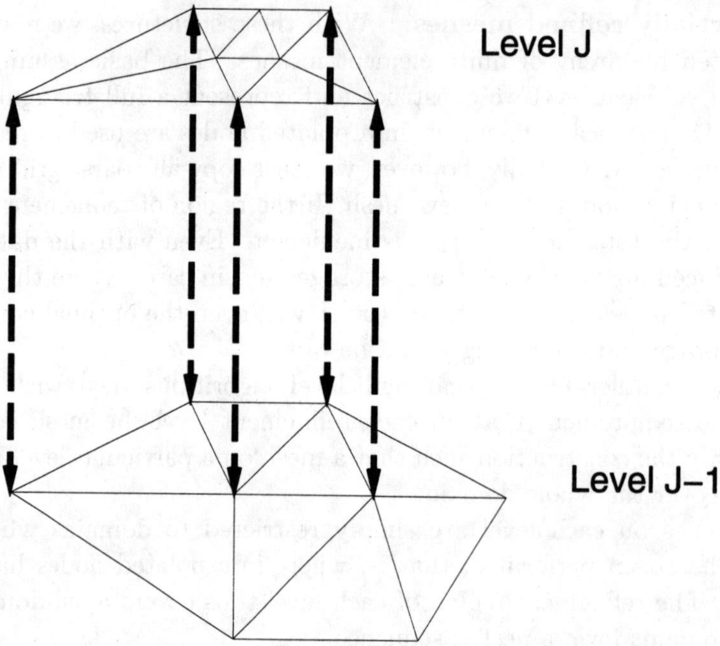

FIG. 4.22. *Hierarchy of two grid levels with partially refined mesh.*

triangulation, but only those lying in a refinement region, that is, $\tilde{\mathcal{N}}^J$ and $\tilde{\mathcal{E}}^J$. The operations on level J are naturally only executed on $\tilde{\mathcal{N}}$, $\tilde{\mathcal{E}}$, and $\tilde{\mathcal{T}}$. In the literature this is referred to as local relaxation. The full finite element mesh for the finer level need never be constructed explicitly.

Within the processing of a level, J, the nodes lying on the boundary of the refinement region play a special role. We call $\partial\Omega^J$ the *interface* of level J, and the nodes on $\partial\Omega^J$ *interface nodes.*

For the processing of a level the interface nodes must be treated as boundary nodes, because they do not have a complete set of neighbors. Therefore, we cannot calculate a residual or relax at the interface points. They are, however, needed to provide data for certain interior nodes of the level, when the residual or a relaxation is to be calculated there. The interface nodes will therefore get some characteristic of *constrained nodes*, but their interaction with coarser grids will be different from points on a Dirichlet boundary. In contrast to Dirichlet boundary points, an interface point will interact with coarser grids, obtaining its value from there, and projecting residuals to its coarse grid parent. These mechanisms are equivalent to the bordered patch interfaces of the FAC method; see McCormick [57]. The functionality required at the patch interfaces leads to an additional node type, the *guard nodes* (see §4.4), whose purpose is to fence in refinement patches. The same mechanisms can also be used for patch interfaces with nonconforming nodes. The required processing here is described in McCormick [57] as *bordered patches.*

The partially refined meshes can be used as the basis for implementing the virtual global meshes introduced in §4.4. Ω^J is the subdomain containing the live nodes and where computations must be performed. The virtual global grid technique implies that the patch interfaces have nonconforming nodes, that is, the guards, so that the implementation of partially refined meshes must be modified at the patch interfaces.

4.5.7. Characteristic sets. For the implementation of the FAMe and the *multilevel adaptive iteration* (MLAI) we use the data structure *set of nodes*; see Chapter 3 and Rüde [88], [91]. These sets are used to *localize* the multilevel processing. They are the basis for implementing error estimation, mesh refinement, and efficient solution.

Thus the efficient implementation of sets is the key to making these methods practical. We must distinguish between different types of sets. Some sets appearing already in the core data structure are Conn(N) and Neigh(N) for all nodes N. These sets must be represented implicitly in the matrix data structure, and indeed, for sparse matrices it is important that they not be trivially implemented, but, e.g., as lists of indices or pointers that support fast sparse matrix multiplication routines. Lists are a suitable representation of *small sets* in our algorithm. Typically, Conn(N) is a small set, because its cardinality is assumed to be much smaller than the system size. We assume that searching a list to determine whether a node is a member of a *small set* is an $O(1)$ operation.

Additionally, the FAMe uses sets to determine where the basic multigrid operations must be executed. As described in Chapter 3 each of the processes *relaxation*, *hierarchical transformation*, *determination of the* τ*-term*, *restriction*, *interpolation*, and *correction* are triggered by characteristic sets that determine where a recalculation of the corresponding values is necessary.

These sets are different, because they vary in size, having close to n elements at some times, but only few elements at other times. A representation of these sets as lists would not allow a fast implementation of the update processes like $\tilde{S} \leftarrow \tilde{S}\backslash\{i\}$ or $\tilde{S} \leftarrow \tilde{S}\cup\text{Conn}(i)$. These must be calculated in $O(1)$ and $O(|\text{Conn}(i)|)$ work, respectively. A straightforward list representation for the sets has the problem of avoiding multiple copies of an index. This generally requires searching and/or sorting techniques, which cannot be implemented within the required complexity bounds.

The alternative representation of sets is by Boolean arrays where each Boolean value indicates whether an element is a member of the set. The Boolean arrays can in turn be implemented as flags for the unknowns. Clearly, this data structure is not suitable for the *small sets* like Conn(i), because each object needs $O(n)$ storage independent of the number of its elements. They are suitable for implementing the update operations for the large sets efficiently, because the updates can be implemented by modifying one or Conn(i) flags, respectively. The problem with the Boolean array representation is operations of type "pick $i \in \tilde{S}$" that may require up to n tests of flags, until an element in the set is found. This is unacceptable, because it contradicts the very reason for introducing the sets in the first place. These sets are used as a device to avoid looping through the full mesh data structure, when an operation need only be performed in a subdomain. Instead of limiting the basic relaxation to the elements in $\tilde{S}$ only, we'd have to loop through all elements testing the flags to find the next element to be relaxed.

The appropriate technique for implementing these large sets is a combination of lists and Boolean arrays. The elements of the large sets are stored in lists of indices or pointers, and each node provides flags indicating whether it is currently in any of these sets. Thus the "pick" operation can be implemented by taking the first element in the list. This is of course an $O(1)$ operation. The updates like $\tilde{S} \leftarrow \tilde{S} \cup \text{Conn}(i)$ can also be implemented efficiently by setting the corresponding flags and appending the indices to the list, if necessary. The removal of an element of the form $\tilde{S} \leftarrow \tilde{S} \setminus \{i\}$ can also be implemented with $O(1)$ operations if the pick operation always selects the first element in the list, and the indices are stored cyclically in an array or a linked list.

Finally, we note that the more complicated operations like

$$\tilde{S}_k \leftarrow \tilde{S}_k \cup A_k I_{k+1}^k V_{k+1},$$

as they occur in the MLAI variant of the FAMe solver (see §3.4 and Rüde [88]), can be implemented based on the simple operations discussed above. For this algorithm the projections $I_{k_1}^{k_2}$ must be sparse and must be stored with sparse matrix techniques, such that $I_{k_1}^{k_2}(i)$ is available as a list of indices.

This is accomplished quite naturally if the interpolated node data structure is extended by information on how the interpolation is to be performed. For the case of triangular elements an interpolation is based on forming the average of two coarse grid values, so that each interpolated node must directly or indirectly contain references to its two interpolators on the coarser grid. In practice, an intermediate large set data structure (with list and flags) may be helpful for holding the intermediate result $I_k^{k+1} V_{k+1}$.

Summing up, the techniques discussed in this section suffice for an implementation of the adaptive algorithms satisfying our efficiency goal. The organizational work is at most proportional to the numerical work performed. The techniques have been applied successfully to implement the prototype solver of Rüde [91].

4.6. Implementation using C++

The data structures have so far been discussed in an abstract setting. In this section we will discuss the practical implementation using C++ and its advanced features. C++ is an extension of the C programming language that provides an efficient implementation of object-oriented programming techniques; see Stroustrup [103] and Bause and Tölle [14].

```
class BasicNode : public Elem {
public:
        BasicNode();
        Boolean HasEdges();
        void InsertEdge(SemiEdge*);
        void Connect(BasicNode*, real );
        Boolean RemoveEdge(SemiEdge*);
        SemiEdge* FindEdgeTo(BasicNode*);
            ...
        friend class SemiEdgeIterator;
}
```

FIG. 4.23. *Basic node data type definition.*

4.6.1. Node classes. C++ offers *classes* to implement abstract data types. A code fragment for the definition of the base node class is shown in Fig. 4.23. This class corresponds to the top class in the hierarchy of Fig. 4.21. This class only implements the topological features of a node in a single level mesh.

The program fragment defines the functionality of the interface to the data type *BasicNode.* It contains functions to manipulate the edges and connections associated with a node of the mesh. This data type does not yet have any geometric or algebraic capabilities.

In the class definition there is no code for the functions given; only their functionality is defined for users of this module. The implementation of the BasicNode is hidden. This allows the internal representation to change without affecting the modules built on top of the basic module.

The first line of the class definition states that *BasicNode* is a derived class of *Elem*, the class that can be a member of a linked list. The list module is a still more basic program unit defining a list data type. Deriving *BasicNode* from the abstract *Elem* type gives it all list-handling capabilities. This mechanism is called *inheritance* and is an important tool for object-oriented modularization and program development. The function declaration for BasicNode() declares a *constructor*, containing the code for the allocation of dynamic storage for a single node. The other member functions of the BasicNode class are designed as an interface to manipulate the topological structure of a mesh.

With the inheritance mechanism, we extend the *BasicNode* type to the other types of the diagram in Fig. 4.21. This way a node is extended to handle geometric and algebraic information and to support the operations necessary in the multilevel context. Figure 4.24 shows a code fragment taken from the definition of the *InitialNode* data type.

It is derived from the *BasicNode* class so that it has all the capabilities already defined in the *BasicNode*, however, it has additional routines to manipulate a geometric location and an instance of the solution value. Note

```
class InitialNode: public BasicNode{
public:
        InitialNode(Wpoint, real val= 0.0);
        Wpoint GetLocation();
        real GetValue();
        void SetValue(real);
              ...
}
```

FIG. 4.24. *Initial node data type definition.*

that the initialization routine (the constructor) requires a parameter of type *Wpoint* that represents a location in the solution domain. The *Wpoint* data type is declared in still another module. By exchanging this module, we can, for example, change from two-dimensional to three-dimensional domains.

The interface to *InitialNode* allows us to set the location of a node only when the node is created. The constructor has a parameter of type *Wpoint*, so that an initial node can only be created when its location is specified. The second argument of the constructor sets the initial value and has a default value of "0," and can therefore be omitted. With this class definition InitialNodes may now be manipulated just like built-in data types; in particular, local variables of type InitialNode may be declared. These InitialNode instances will follow the usual scoping and lifetime rules of local variables. With the *new* operator, InitialNodes may also be generated on the heap so that dynamic structures can be constructed with them.

When an *InitialNode* has been created, its location may be requested by a call to *GetLocation*, however, the location cannot be changed because the interface does not provide an operation for this. Once a node has been created, its location is fixed. The numerical value, in contrast, may be either read or written, because this is required within the iterative solution process.

In this style a hierarchical system of classes may be built that represents the features of the abstract data types discussed in §4.2. Among the classes to be designed are other classes of nodes, like *interpolated* and *replicated* nodes, but also a class *Mesh* that allows access to all the nodes of a finite element mesh.

4.6.2. Efficiency for special cases requiring limited mesh flexibility. Continuing this process, we will finally derive a piece of software that implements general finite element meshes and multilevel solvers. As discussed in §4.3, such general structures are necessarily inefficient in situations where the full flexibility is not needed. At present it is unclear whether the unstructured meshes are necessary at all. They have great structural disadvantages that may not be compensated for by their additional flexibility. In any case, it is unfortunate that a complicated, adaptive multilevel program for unstructured meshes will not have a chance to be an efficient solver for, say, Laplace's equation in a rectangular domain as well. Ideally, the same program should be a reasonable solver for all situations. Though we do not expect that a generic solver will have the same efficiency as a program tuned for a special case, we are also unwilling to accept that it should have an overhead that can be of orders of magnitude.

In the following, we will develop some ideas for how this problem can be alleviated. The key to such generic software is modularization—modularization beyond the usual one, however.

The main overhead in a general mesh is the complexity inherent in the access to its nodes, edges, etc. Thus our goal must be to make the mesh structure itself a module that can be replaced by a simpler one to trade efficiency for flexibility, if appropriate.

Keeping the mesh class abstract, we have the opportunity to implement the different mesh structures of §4.3 in different variants of this module. Thus, by simply exchanging this module, we can change from restricted but efficient uniform grids to the maximal flexibility of unstructured meshes.

For this step to work, we must find an abstract interface to the mesh type. The main problems here are the access routines. For a uniform grid we wish to represent the mesh simply by a (two-)dimensional array, while unstructured meshes will require indirect access through a linked list. Correspondingly, the access structure for the uniform grid consists of two nested loops. For nodes in unstructured meshes the corresponding structure is a while loop or a recursion.

To keep the package modular and flexible, the access function must be exchangeable. In a precursor version of the FAMe package (see Rüde [91]) that was still implemented in C, this had been accomplished by an involved macro package that implemented generic access routines, like a *ForAllNodes* construct. The macro *ForAllNodes* could be expanded to either simple loops for the uniform case or an iteration through a linked list for the unstructured case. This abstraction is important, because the access structures occur throughout the program, making changes very difficult.

An alternative technique is used by Hemker, van der Maarel, and Everaars [45] for the BASIS data structure. Here the access is handled by a routine *scan* that loops through all elements and applies a function *doit* that is a parameter to the *scan* routine. Unfortunately, this technique has several disadvantages. For each loop an extra routine must be written making the code less readable. Furthermore, each operation costs the overhead of a function call, and finally the technique has severe limitations, because operations like double loops over the mesh cannot be programmed with the functions provided.

With the C++ language a better abstraction is possible using the technique of *iterator classes*; see Stroustrup [103] and Dewhurst and Stark [36]. Abstract iterators are also used to access the edges (connections) from a node; see the *friend* declaration in the *BasicNode* class of Fig. 4.23. With these iterators a loop through all the nodes is written as shown in Fig. 4.25.

```
FEMNet *mesh
    ...
NodeIterator i(mesh); // declaration of iterator
BasicNode* node;
while(node= i.GetElem()){ // i loops though the mesh
        "do something with the node"
}
```

FIG. 4.25. *Application of abstract node iterator.*

The declaration of an iterator has a parameter of type *FEMNet* that initializes the iterator. The *GetElem* operation advances the iterator by one node and returns 0 at termination. The advantage of this technique is that the code to access the nodes stays the same, no matter what type of mesh *FEMNet* represents. Only the iterator class must be modified appropriately to match the mesh structure. If the iterator class is implemented well (using function inlining[4]) this can be accomplished with little or no overhead.

A truly modular object-oriented design will even go one step further. The primary function of the *Mesh* class is that of a *container* of nodes, where container is a generic concept that includes arrays, lists, sets, etc.

For unstructured meshes the list class can be used as a suitable container class. As mentioned above, the BasicNode class should therefore be derived (inherited) from the *Elem* class, the Mesh class from the *List* class, and finally, the NodeIterator class from the generic *ElemIterator* provided with the list module.

If we wish to switch to uniform meshes, we can now just replace the list, elem, and list-iterator classes with a suitable implementation of two-dimensional arrays providing an equivalent abstract interface. Thus we can use the same solver routines with different mesh structures and recover most of the computational efficiency of simpler data structures.

In this fashion all the mesh structures of §4.3 could be accommodated. Following this implementation guideline we therefore derive truly abstract solvers that do not depend on a particular mesh data structure, but only on the abstract interface to generic meshes.

4.6.3. Virtual overloaded functions. When programming the multilevel components one finds that in loops through the mesh the operation to be performed for each node usually depends on the node type, e.g., whether it is a replicated or an interpolated node. Therefore, we frequently have an *if* or *case* statement within the loop. In C++ we can deal with this better by using *operator overloading.* Operator overloading permits the use of the same

[4]Function *inlining* is an optimization technique where the code for a function is copied to the point where the function is called. This avoids the overhead of subroutine calls and simplifies further automatic code optimization by the compiler.

function name for different operations if the type of the objects is different. If these overloaded functions are declared *virtual*, the binding of the function name to the actual routine is done at run time, depending on the type of the object addressed. This typical object-oriented technique is equivalent to a classical function call within a case statement distinguishing the different situations. In our context, using the overloaded virtual function has great advantages for the modularity and extensibility of the program.

As an example, consider the interpolation routine that transports the solution value from the coarser level to the current one. This operation must perform different actions for interpolated and replicated nodes. In C++ both routines may have the same name and the routine to be called for each node is automatically determined by the type of the node. If this is done dynamically at program run time, the corresponding functions are called *virtual functions.*

A call to a virtual function is more expensive than a call to a statically bound function. Therefore, C++ leaves the decision of which calling mechanism should be used to the programmer. Functions must be declared *virtual* explicitly, if the binding at run time is desired. Though virtual functions cause some overhead, they are a convenience in program development because they help to reduce the code complexity. They can be used to avoid the case- or if-constructs otherwise necessary to select the correct functions for each node type. With operator overloading and virtual functions this distinction is made implicitly in the definition of the class system. Equally important, they improve the extensibility of the code because new functionality can be added more easily by extending the class system.

4.6.4. The role of C++. With these examples we have outlined how data abstraction and object-oriented techniques in general, and the language C++ in particular, can be used to develop an efficient and modular implementation of the abstract data types for finite element meshes. With these programming techniques the abstract structure can be implemented so that maximal modularity, flexibility, and extensibility are achieved. If designed carefully and implemented appropriately, the resulting program will also be efficient.

We wish to point out that none of the software engineering techniques presented above is new per se. Such modularization techniques are consequences of an abstraction that is more or less natural if the software development is liberated from the restrictions of the presently used languages. Note, however, that with a disciplined programming style many of the techniques can also be used with classical languages, especially when preprocessors are used. For examples of what can be done even with plain FORTRAN, but also of what the limitations of such an approach are, see Jänsch, Rüde, and Schnepper [47].

As demonstrated in the above examples, the C++ extension of C is well suited to implement such an advanced software architecture directly. Though other languages offer similar functionality, C++ seems to be the first choice at present, because of its inherent efficiency (if used appropriately) and also because of its availability, tools, support, etc. It should be remarked, however, that C++ has several problems, some inherited from its ancestor C, some due to the fact that it is a complicated and rich language whose effective use requires experience and careful programming.

References

[1] I. ADLER, *Zur Spezifikation und Implementierung von finiten Element Netzgeneratoren*, Diplomarbeit, Institut für Informatik, TU München, May 1987.

[2] O. AXELSSON AND V. BARKER, *Finite element solutions of boundary value problems*, Academic Press, New York, 1984.

[3] I. BABUŠKA AND A. AZIZ, *On the angle condition in the finite element method*, SIAM J. Numer. Anal., 13 (1976), pp. 214–226.

[4] I. BABUŠKA, A. MILLER, AND M. VOGELIUS, *Adaptive methods and error estimation for elliptic problems in structural mechanics*, in Adaptive Computational Methods for Partial Differential Equations, I. Babuška, J. Chandra, and J. Flaherty, eds., Society for Industrial and Applied Mathematics, Philadelphia, 1983.

[5] I. BABUŠKA, B. A. SZABO, AND I. KATZ, *The p-version of the finite element method*, SIAM J. Numer. Anal., 18 (1981), pp. 515–545.

[6] D. BAI AND A. BRANDT, *Local mesh refinement multilevel techniques*, SIAM J. Sci. Statist. Comput., 8 (1987), pp. 109–134.

[7] R. BANK AND M. BENBOURENANE, *The hierarchical basis multigrid method for convection-diffusion equations*, preprint, Department of Mathematics, University of California at San Diego, 1991.

[8] R. BANK, T. DUPONT, AND H. YSERENTANT, *The hierarchical basis multigrid method*, Numer. Math., 52 (1988), pp. 427–458.

[9] R. BANK AND A. WEISER, *Some a-posteriori estimators for elliptic partial differential equations*, Math. Comp., 44 (1985), pp. 283–301.

[10] R. E. BANK, *PLTMG: A Software Package for Solving Elliptic Partial Differential Equations*, Frontiers in Applied Mathematics, Vol. 7, Society for Industrial and Applied Mathematics, Philadelphia, 1990.

[11] E. BÄNSCH, *Local mesh refinement in* 2 *and* 3 *dimensions*, IMPACT of Computing in Science and Engineering, 3 (1991), pp. 181–191.

[12] P. BASTIAN, *Locally refined solution of unsymmetric and nonlinear problems*, in Incomplete Decompositions: Proceedings of the Eigth GAMM-Seminar, Notes in Numerical Fluid Mechanics, Vol. 41, Vieweg-Verlag, Braunschweig, 1992, pp. 12–21.

[13] F. BAUER AND H. WÖSSNER, *Algorithmische Sprache und Programmentwicklung*, 2. Auflage, Springer-Verlag, 1984. English translation: Algorithmic Language and Program Development.

[14] F. BAUSE AND W. TÖLLE, *C++ für Programmierer*, Vieweg-Verlag, Braunschweig, 1990.

[15] O. BESOV, L. KUDRAYAVTSEV, P. LIZORKIN, AND S. NIKOL'SKIĬ, *Investigations in the theory of spaces of differentiable functions of several variables*, Proceedings

of the Steklov Institute of Mathematics, St. Petersburg, Russia, 1990.

[16] H. BLUM, *Asymptotic error expansion and defect correction in the finite element method*, Habilitationsschrift, Universität Heidelberg, 1991.

[17] H. BLUM, Q. LIN, AND R. RANNACHER, *Asymptotic error expansions and Richardson extrapolation for linear finite elements*, Numer. Math., 49 (1986), pp. 11–37.

[18] H. BLUM AND R. RANNACHER, *Extrapolation techniques for reducing the pollution effect of reentrant corners in the finite element method*, Numer. Math., 52 (1988), pp. 539–564.

[19] T. BONK AND U. RÜDE, *Performance analysis and optimization of numerically intensive programs*, SFB Bericht 342/26/92 A, Institut für Informatik, TU München, November 1992.

[20] F. BORNEMANN AND H. YSERENTANT, *A basic norm equivalence for the theory of multilevel methods*, Numer. Math., 64 (1993), pp. 455–476.

[21] D. BRAESS, *Finite Elemente*, Springer-Verlag, Berlin, 1992.

[22] D. BRAESS AND W. HACKBUSCH, *A new convergence proof for the multigrid method including the V-cycle*, SIAM J. Numer. Anal., 20 (1983), pp. 967–975.

[23] J. BRAMBLE, J. PASCIAK, J. WANG, AND J. XU, *Convergence estimates for multigrid algorithms without regularity assumptions*, Report BNL-44512, Brookhaven National Laboratory, New York, August 1989.

[24] ——, *Convergence estimates for product iterative methods with applications to domain decomposition and multigrid*, Report BNL-44511, Brookhaven National Laboratory, New York, August 1989.

[25] J. BRAMBLE, J. PASCIAK, AND J. XU, *Parallel multilevel preconditioners*, Math. Comp., 31 (1990), pp. 333–390.

[26] J. H. BRAMBLE, J. E. PASCIAK, AND J. XU, *The analysis of multigrid algorithms with nonnested spaces or noninherited quadratic forms*, Math. Comp., 56 (1991), pp. 1–34.

[27] A. BRANDT, *Multi-level adaptive solutions to boundary value problems*, Math. Comp., 31 (1977), pp. 333–390.

[28] ——, *Multigrid techniques:* 1984 *guide with applications to fluid dynamics*, GMD Studien, Vol. 85, 1984.

[29] ——, *Algebraic multigrid theory: The symmetric case*, Appl. Math. Comp., 19 (1986), pp. 23–65.

[30] ——, *Rigorous local mode analysis of multigrid*, in Preliminary Proceedings of the Fourth Copper Mountain Conference on Multigrid Methods, University of Colorado at Denver, April 9–14, 1989, S. McCormick, ed.

[31] A. BRANDT AND D. OPHIR, *Gridpack*, in Proceedings IFIP Conference on PDE Software, Solderkoping, Sweden, 1983.

[32] W. L. BRIGGS, *A Multigrid Tutorial*, Society for Industrial and Applied Mathematics, Philadelphia, 1987.

[33] P. CIARLET, *The Finite Element Method for Elliptic Problems*, North-Holland, Amsterdam, New York, London, 1978.

[34] W. DAHMEN AND A. KUNOTH, *Multilevel preconditioning*, Numer. Math., 63 (1992), pp. 315–344.

[35] P. DEUFLHARD, P. LEINEN, AND H. YSERENTANT, *Concepts of an adaptive hierarchical finite element code*, IMPACT of Computing in Science and Engineering, 1 (1989), pp. 3–35.

[36] S. DEWHURST AND K. STARK, *Programming in C++*, Prentice Hall, Englewood Cliffs, NJ, 1989.

[37] M. DRYJA AND O. WIDLUND, *An additive variant of the Schwarz alternating method for the case of many subregions*, Tech. Report 339 and Ultracomputer Note 131, Department of Computer Science, Courant Institute, 1987.

[38] ——, *Multilevel additive methods for elliptic finite element problems*, in Parallel Algorithms for Partial Differential Equations, Proceedings of the Sixth GAMM-Seminar, Kiel, January 19–21, 1990, W. Hackbusch, ed., Vieweg-Verlag, Braunschweig, 1991.

[39] A. GEORGE AND W. LIU, *Computer Solution of Large Sparse Positive Definite Systems*, Prentice Hall, Englewood Cliffs, NJ, 1981.

[40] M. GRIEBEL, *Zur Lösung von Finite-Differenzen- und Finite-Element-Gleichungen mittels der Hierarchischen-Transformations-Mehrgitter-Methode*, SFB Bericht 342/4/90 A, Institut für Informatik, TU München, February 1990.

[41] ——, *Multilevel algorithms considered as iterative methods on indefinite systems*, SFB Bericht 342/29/91 A, Institut für Informatik, TU München, October 1991; see also Preliminary Proceedings of the Second Copper Mountain Conference on Iterative Methods.

[42] ——, *Grid- and point-oriented multilevel algorithms*, SFB Bericht 342/14/92 A, Institut für Informatik, TU München, July 1992.

[43] W. HACKBUSCH, *Multigrid Methods and Applications*, Springer-Verlag, Berlin, 1985.

[44] ——, *Theorie und Numerik elliptischer Differentialgleichungen*, Teubner Studienbücher, Stuttgart, 1986.

[45] P. HEMKER, H. VAN DER MAAREL, AND C. EVERAARS, *BASIS: A data structure for adaptive multigrid computations*, Report NM-R9014, Department of Numerical Mathematics, Centrum for Wiskunde en Informatica, Amsterdam, August 1990.

[46] R. HEMPEL AND H. RITZDORF, *The GMD communications library for grid-oriented problems*, Arbeitspapiere der GMD 589, Gesellschaft für Mathematik und Datenverarbeitung, St. Augustin, 1991.

[47] C. R. JÄNSCH, U. RÜDE, AND K. SCHNEPPER, *Macro expansion, a tool for the systematic development of scientific software*, Bericht I-8814, Institut für Informatik, TU München, November 1988.

[48] H. JOHNEN AND K. SCHERER, *On the equivalence of the K-functional and the moduli of continuity and some applications*, Lecture Notes in Mathematics 571, Springer-Verlag, New York, 1977, pp. 119–140.

[49] P. LEINEN, *Ein schneller adaptiver Löser für elliptische Randwertprobleme auf Seriell- und Parallelrechnern*, Dissertation, Universität Dortmund, 1990.

[50] P. LEINEN AND H. YSERENTANT, Private communication, 1991.

[51] M. LEMKE AND D. QUINLAN, *P++, a C++ virtual shared grids based programming environment for architecture-independent development of structured grid applications*, Arbeitspapiere der GMD 611, Gesellschaft für Mathematik und Datenverarbeitung, St. Augustin, Germany, February 1992.

[52] R. LÖHNER AND K. MORGAN, *Unstructured multigrid methods*, in Multigrid Methods: Special Topics and Applications, Papers presented at the Second European Conference on Multigrid Methods, Cologne, October 1–4, 1985, U. Trottenberg and W. Hackbusch, eds., GMD Studien, Vol. 110, 1986.

[53] D. LUENBERGER, *Introduction to linear and nonlinear programming*, Addison-Wesley, Reading, MA, 1973.

[54] G. MARCHUK AND V. SHAIDUROV, *Difference Methods and Their Extrapolations*, Springer-Verlag, New York, 1983.

[55] O. MCBRYAN, P. FREDERICKSON, J. LINDEN, A. SCHÜLLER, K. SOLCHENBACH,

K. Stüben, C.-A. Thole, and U. Trottenberg, *Multigrid methods on parallel computers—a survey of recent developments*, IMPACT of Computing in Science and Engineering, 3 (1991), pp. 1–75.

[56] S. McCormick, ed., *Multigrid Methods*, Frontiers in Applied Mathematics, Vol. 3, Society for Industrial and Applied Mathematics, Philadelphia, 1987.

[57] S. McCormick, *Multilevel Adaptive Methods for Partial Differential Equations*, Frontiers in Applied Mathematics, Vol. 6, Society for Industrial and Applied Mathematics, Philadelphia, 1989.

[58] ——, *Multilevel Projection Methods for Partial Differential Equations*, CBMS-NSF Regional Conference Series in Applied Mathematics, Society for Industrial and Applied Mathematics, Philadelphia, 1992.

[59] S. McCormick and U. Rüde, *On local refinement higher order methods for elliptic partial differential equations*, Internat. J. High Speed Comput., 2 (1990), pp. 311–334; also available as TU-Bericht I-9034, Institut für Informatik, TU München, September 1990.

[60] S. McCormick and J. Ruge, *Multigrid methods for variational problems*, SIAM J. Numer. Anal., 19 (1982), pp. 924–927.

[61] ——, *Unigrid for multigrid simulation*, Math. Comp., 41 (1983), pp. 43–62.

[62] S. McCormick and J. Thomas, *The fast adaptive composite grid (FAC) method for elliptic equations*, Math. Comp., 46 (1986), pp. 439–456.

[63] V. Mikulinsky, *Multigrid Treatment of Boundary and Free-Boundary Conditions*, Ph.D. thesis, The Weizmann Institute of Science, Rehovot, Israel, 1992.

[64] W. Mitchell, *Adaptive refinement for arbitrary finite element spaces with hierarchical bases*, J. Comp. Appl. Math., 36 (1991), pp. 65–78.

[65] ——, *Optimal multilevel iterative methods for adaptive grids*, SIAM J. Sci. Statist. Comput., 13 (1992), pp. 146–167; also in Proceedings of the First Copper Mountain Conference on Iterative Methods, April 1–5, 1990, T. Manteuffel, ed.

[66] S. Nikol'skiĭ, *Approximation of Functions of Several Variables and Imbedding Theorems*, Springer-Verlag, New York, 1975 (translated from Russian).

[67] A. Ostrowski, *Determinanten mit überwiegender Hauptdiagonale und die absolute Konvergenz von linearen Iterationsprozessen*, Comment. Math. Helv., 30 (1956), pp. 175–210.

[68] P. Oswald, *On C^1-interpolating hierarchical spline bases*, Tech. Report N/89/19, Friedrich Schiller Universität, Jena, July 1989.

[69] ——, *On estimates for hierarchic basis representations of finite element functions*, Tech. Report N/89/16, Friedrich Schiller Universität, Jena, June 1989.

[70] ——, *On function spaces related to finite element approximation theory*, Z. Anal. Anwendungen, 9 (1990), pp. 43–64.

[71] ——, *On the degree of nonlinear spline approximation in Besov-Sobolev spaces*, J. Approxim. Theory, 61 (1990), pp. 131–157.

[72] ——, *On discrete norm estimates related to multilevel preconditioners in the finite element method*, in Proceedings of International Conference on Constructive Theory of Functions, Varna 1991, Bulgarian Academy of Science, Sofia, K. Ivanov and B. Sendov, 1992, pp. 203–241.

[73] ——, *Preconditioners for discretizations of the biharmonic equation by rectangular finite elements*, Forschungsergebnisse Math/91/03, Friedrich Schiller Universität, Jena, November 1991.

[74] ——, *Two remarks on multilevel preconditioners*, Forschungsergebnisse Math/91/1, Friedrich Schiller Universität, Jena, 1991.

[75] P. OSWALD, *Norm equivalencies and multilevel Schwarz preconditioning for variational problems*, Forschungsergebnisse Math/92/01, Friedrich Schiller Universität, Jena, January 1992.

[76] ——, *Stable splittings of Sobolev spaces and fast solution of variational problems*, Forschungsergebnisse Math/92/05, Friedrich Schiller Universität, Jena, May 1992.

[77] L. PAVARINO, *Some Schwarz algorithms for the p-version of the finite element method*, Tech. Report 614, Department of Computer Science, Courant Institute, New York University, 1992.

[78] A. QUARTERONI, *Domain decomposition and parallel processing for the numerical solution of partial differential equations*, Surv. Math. Ind., 1 (1991), pp. 75–118.

[79] H. REGLER AND U. RÜDE, *Layout optimization with algebraic multigrid methods (AMG)*, in Proceedings of the Sixth Copper Mountain Conference on Multigrid Methods, Copper Mountain, April 4–9, 1993, Conference Publication, NASA, 1993.

[80] M. RIVARA, *Algorithms for refining triangular grids suitable for adaptive and multigrid techniques*, Internat. J. Numer. Methods Engrg., 20 (1984), pp. 745–756.

[81] ——, *Design and data structure of fully adaptive, multigrid, finite-element software*, ACM Trans. Math. Software, 10 (1984), pp. 242–264.

[82] R. ROITZSCH, *Kaskade user's manual*, Tech. Report TR 89-4, Konrad-Zuse-Zentrum für Informationstechnik, Berlin, August 1989.

[83] U. RÜDE, *Multiple tau-extrapolation for multigrid methods*, Bericht I-8701, Institut für Informatik, TU München, January 1987.

[84] ——, *On the accurate computation of singular solutions of Laplace's and Poisson's equation*, in Multigrid Methods: Theory, Applications, Supercomputing: Proceedings of the Third Copper Mountain Conference on Multigrid Methods, April 5–10, 1987, S. McCormick, ed., Marcel Dekker, New York, 1988.

[85] ——, *Extrapolation and related techniques for solving elliptic equations*, Bericht I-9135, Institut für Informatik, TU München, September 1991.

[86] ——, *Data structures for multilevel adaptive methods and iterative solvers*, Bericht I-9217, Institut für Informatik, TU München, May 1992.

[87] ——, *The hierarchical basis extrapolation method*, SIAM J. Sci. Statist. Comput., 13 (1992), pp. 307–318; also in Proceedings of the First Copper Mountain Conference on Iterative Methods, April 1–5, 1990, T. Manteuffel, ed.

[88] ——, *On the multilevel adaptive iterative method*, in Preliminary Proceedings of the Second Copper Mountain Conference on Iterative Methods, April 9–14, 1992, T. Manteuffel, ed.; SIAM J. Sci. Comput., 15 (1994), to appear.

[89] ——, *Data abstraction techniques for multilevel algorithms*, in Proceedings of the GAMM-Seminar on Multigrid Methods, in Gosen, Germany, Sept. 21–25, 1992; Report 5, Institut für Angewandte Analysis und Stochastik, Berlin, 1993. ISSN 0942–9077.

[90] ——, *Extrapolation techniques for constructing higher order finite element methods*, Bericht I-9304, Institut für Informatik, TU München, 1993; SIAM J. Numer. Anal., submitted.

[91] ——, *Fully adaptive multigrid methods*, SIAM J. Numer. Anal., 30 (1993), pp. 230–248.

[92] U. RÜDE AND C. ZENGER, *On the treatment of singularities in the finite element method*, Bericht I-9220, Institut für Informatik, TU München, August 1992.

[93] H. SCHWARZ, *Methode der Finiten Elemente*, 3. Auflage, Teubner Studienbücher Mathematik, Teubner, Stuttgart, 1991.

[94] L. SEIDEL, *Über ein Verfahren, die Gleichungen, auf welche die Methode der kleinsten Quadrate führt, sowie lineare Gleichungen überhaupt, durch successive*

Annäherung zu lösen, Abh. Bayer. Akademie der Wiss. Math. Nat. Kl. 11, 1874.

[95] E. SEWELL, *A finite element program with automatic user-controlled mesh grading*, in Advances in Computer Methods for Partial Differential Equations III, R. Vichnevetsky and R. Stapleman, eds., IMACS, New Brunswick, NJ, 1979, pp. 8–10.

[96] R. SOUTHWELL, *Stress-calculation in frameworks by the method of systematic relaxation of constraints*, Parts I, II, Proc. Roy. Soc. (A), (1935), pp. 56–95.

[97] ——, *Stress-calculation in frameworks by the method of "systematic relaxation of constraints,"* Part III, Proc. Roy. Soc. (A), (1935), pp. 41–76.

[98] ——, *Relaxation Methods in Engineering Science—a Treatise in Approximate Computation*, Oxford University Press, 1940.

[99] ——, *Relaxation Methods in Theoretical Physics*, Clarendon Press, Oxford, 1946.

[100] M. STARK, *Spezifikation und Implementierung einer Datenstruktur für dreidimensionale finite Elemente*, Diplomarbeit, Institut für Informatik, TU München, May 1989.

[101] R. STEVENSON, *On the Validity of Local Mode Analysis of Multi-Grid Methods*, Ph.D. thesis, University of Utrecht, the Netherlands, 1990.

[102] G. STRANG AND G. FIX, *An Analysis of the Finite Element Method*, Prentice-Hall, Englewood Cliffs, NJ, 1973.

[103] B. STROUSTRUP, *The C++ Programming Language*, Second Ed., Addison-Wesley, Reading, MA, 1991.

[104] H. TRIEBEL, *Interpolation Theory, Function Spaces, Differential Operators*, North Holland, Amsterdam, 1978.

[105] ——, *Theory of Function Spaces*, Monographs in Mathematics, Vol. 78, Birkhäuser-Verlag, Basel, 1983.

[106] R. VERFÜHRT, *A review of a posteriori error estimation and adaptive mesh-refinement techniques*, Lecture Notes of a Compact Seminar at the TU Magdeburg, June 2–4, 1993; also Report, Universität Zürich.

[107] O. B. WIDLUND, *Optimal iterative refinement methods*, in Domain Decomposition Methods, T. F. Chan, R. Glowinsky, J. Périaux, and O. B. Widlund, eds., Society for Industrial and Applied Mathematics, Philadelphia, 1989.

[108] J. XU, *Theory of multilevel methods*, Report AM48, Department of Mathematics, Pennsylvania State University, State College, PA, July 1989.

[109] ——, *Iterative methods by space decomposition and subspace correction: A unifying approach*, Report AM67, Department of Mathematics, Pennsylvania State University, State College, PA 1990.

[110] ——, *A new class of iterative methods for nonselfadjoint or indefinite problems*, SIAM J. Numer. Anal., 29 (1992), pp. 303–319.

[111] H. YSERENTANT, *On the multi-level splitting of finite element spaces*, Numer. Math., 49 (1986), pp. 379–412.

[112] ——, *Two preconditioners based on the multi-level splitting of finite element spaces*, Numer. Math., 58 (1990), pp. 163–184.

[113] ——, *Hierarchical bases*, in Proceedings of the ICIAM 1991, Society for Industrial and Applied Mathematics, 1992.

[114] ——, *Old and new convergence proofs for multigrid methods*, Acta Numerica, (1993), pp. 285–326.

[115] S. ZHANG, *Multilevel iterative techniques*, Report, Department of Mathematics, Pennsylvania State University, State College, PA, 1988.

[116] X. ZHANG, *Multilevel additive Schwarz methods*, Tech. Report 582, Department of Computer Science, Courant Institute, New York University, 1991.

Index